Beyond LEO - Human Health Issues for Deep Space Exploration

Edited by Robert J. Reynolds

Published in London, United Kingdom

IntechOpen

Supporting open minds since 2005

Beyond LEO - Human Health Issues for Deep Space Exploration
http://dx.doi.org/10.5772/intechopen.77436
Edited by Robert J. Reynolds

Contributors
Jean-Pol Frippiat, Sandra Kaminski, Stéphanie Ghislin, Julie Bonnefoy, Gaetano Calcagno, Coralie Fonte, Jeremy Jeandel, Arthur Hand, Maija Mednieks, Lijian Shao, Huihong Zeng, Mengzhen Yue, Satoshi Iwase, Naoki Nishimura, Tadaaki Mano, Kunihiko Tanaka, Robert J. Reynolds, Mark Shelhamer

First published in London, United Kingdom, 2020 by IntechOpen
IntechOpen is the global imprint of INTECHOPEN LIMITED, registered in England and Wales, registration number: 11086078, 7th floor, 10 Lower Thames Street, London, EC3R 6AF, United Kingdom
Printed in Croatia

British Library Cataloguing-in-Publication Data
A catalogue record for this book is available from the British Library

Additional hard and PDF copies can be obtained from orders@intechopen.com

Beyond LEO - Human Health Issues for Deep Space Exploration
Edited by Robert J. Reynolds
p. cm.
Print ISBN 978-1-78985-509-8
Online ISBN 978-1-78985-510-4
eBook (PDF) ISBN 978-1-83880-849-5

For EU product safety concerns:
IN TECH d.o.o., Prolaz Marije Krucifikse Kozulić 3, 51000 Rijeka, Croatia, info@intechopen.com or visit our website at intechopen.com.

Meet the editor

Robert Reynolds is a Consulting Research Scientist for Mortality Research & Consulting, Inc., and a Visiting Data Scientist for the Human Health and Performance Directorate (HHPD) at NASA Johnson Space Center, through the Translational Research Institute for Space Health (TRISH) at the Baylor College of Medicine in Houston. Dr. Reynolds has a background in epidemiology and biostatistics, and is a Professionally Accredited Statistician® by the American Statistical Association. He has published extensively on the mortality of space explorers, including US astronauts, Soviet and Russian cosmonauts, and International Partner astronauts. He is a self-trained health data scientist interested in how advanced analytics and machine learning can be employed to support operational decision making for human health and performance in spaceflight.

Contents

Preface

The first 60 years of manned space exploration has seen great advances in technology and achievement. Not the least of these are the great leaps in knowledge made through observation and experimentation in the biological and medical sciences. Though the United States National Aeronautics and Space Administration (NASA) completed twelve crewed missions to the moon between 1969 and 1972, the vast majority of our knowledge of human physiology and response to space flight comes from extended stays in low Earth orbit on the space station Mir, the Space Shuttle, and the International Space Station.

NASA has expressed its intention to conduct crewed expeditions to the moon in the 2020s, working toward the goal of humans visiting Mars in the 2030s. The extended duration and distance from Earth these missions entail pose a number of new challenges for space agencies seeking to send humans into the abyss and return them safely home to Earth. New knowledge and new technology will be needed to conquer these challenges.

This book presents a small sample of the physiological changes and human health risks that have been observed in low Earth orbit, and that will undoubtedly be magnified with extended exploration operations to deep space. This book presents the evidence to date and offers a glimpse at what will be needed to take humanity further into deep space than ever before.

Robert J. Reynolds, MS MPH PhD
Mortality Research and Consulting Inc.,
Translational Research Institute for Space Health,
Baylor College of Medicine,
USA

Section 1

Introduction

Chapter 1

Introductory Chapter: Research Methods for the Next 60 Years of Space Exploration

Robert J. Reynolds and Mark Shelhamer

1. Introduction

There are many potential health hazards inherent to space travel, and, as the chapters in this book make clear, even after 60 years of human space exploration, much is left to be learned about how to live and work in space. As a result of the diversity of problems that remain to be solved, the scientific methods required to research these issues need to be flexible and varied. This is perhaps most true in our approach to analyzing data and drawing conclusions from them in the context of space medicine.

In a commentary published in the Journal of Applied Physiology, Ploutz-Snyder et al. [1] point out that in the study of exotic topics (such as the physiology and health of space travelers) the available data are often insufficient to satisfy the sample-size requirements for traditional null-hypothesis statistical testing (NHST). They rightly point out that if we hold this as the standard of good research, (i.e., if NHST is our only, or even our preferred, tool for learning from data) we will be forced to abandon whole lines of research. While the authors offer several "approaches for justifying small-n research," even these are attempts to shoehorn small datasets into traditional statistical analysis. This misses the broader (epistemological) point: what is needed in small-n studies is not just a better way to use statistics, but rather other tools which afford the freedom to learn without using statistics at all.

2. The problems of small-n settings

Research on small sample sizes poses a number of challenges. First and foremost is the violation of assumptions that frequentist statistical methods often require in order to be valid. Secondary to this, but inherent in the nature of small samples, is the typical lack of statistical power for detecting differences other than those in low-variance settings or those with dramatic effects. Each of these two challenges can lead to difficulty in interpreting results.

2.1 Violations of frequentist assumptions

Most frequentist statistical analyses follow a familiar pattern: assume the outcome follows a known statistical distribution, then test whether or not the observed data are unusual (unexpected) under the null hypothesis. However, beyond basic goodness-of-fit considerations, such analyses require other assumptions as well, many of which are clearly violated much of the time. Perhaps the most important of these assumptions is that the observations in a given sample are "iid"—independent and identically distributed. When samples are strictly observational (i.e., not from a randomized trial) this assumption is often unwarranted. The implication of violating this assumption can be profound: differential probability of exposure and inequitable distributions of potential confounders can lead to what is known as *confounding by indication*, a subtle form of bias that can lead to misleading or even wholly wrong conclusions.

2.2 Statistical power

In any statistical analysis our strength of conviction for our conclusions is largely dependent on how much data we can observe (sample size), and how consistent our outcomes are within those observed data (variance). In the frequentist statistical context this is reflected in the concept of *statistical power*. Statistical power is defined as the (hypothetical) probability of correctly rejecting the null hypothesis when the null hypothesis is indeed false (and false by a pre-set threshold considered to be of clinical or practical importance). A commonly desired and accepted level of statistical power is 80%. However, it should be noted that even with this level of power, there is a 20% chance of making a Type II error (i.e., incorrectly failing to reject the null hypothesis). Unless the ratio of the standard deviation to the mean (coefficient of variation) is small, the statistical power in small studies is considerably lower than 80%, effectively crippling the ability to confidently draw inferences from the data under this framework.

2.3 Interpretation

Both violation of assumptions and low statistical power can frustrate the drawing of inferences under traditional statistical approaches. If we manage to obtain a statistically significant effect, how should we interpret it given the potential for confounding by indication? If we fail to see any significant effects where we believe we ought to *a priori*, how do we interpret that? Does our assessment of the meaning of such results change with larger or smaller variance in our sample? Under traditional approaches we surrender to the probabilities of committing Type I or Type II errors, and resign ourselves to having learned nothing.

2.4 Preference for errors

The ultimate motive for use of the NHST framework is to reach reasonable conclusions about a population or process from a (large) subsample of it. However, a real yet unintended consequence of the framework is the focus on avoidance of error. The framework itself is centered on the concept of errors in inference: when we can, we design our studies to avoid Type I error while simultaneously trying to limit Type II errors. In so doing we may make these errors—rather than what we might learn from our data—the primary consideration of our scientific activity. It should come as no surprise that when we make avoidance of error our top priority, we fail to learn all we can from our data.

Modern science's focus on Type I error has proven to be particularly troublesome. In our quest to never actively assert a false truth we have no doubt passively allowed many truths to go unspoken. It is obvious that Type I errors can cause harm in medicine if new treatments are adopted that are actually harmful to patients. Less obvious is the harm that may result if research into a truly efficacious treatment is abandoned simply because a p-value was too high. Such harm is every bit as real (and every bit as irreversible) as that done by introducing an ineffective treatment. It is especially troubling in initial exploratory studies and those where data are acquired only with great difficulty or expense.

3. Methodological solutions for research in space medicine

Having seen the problems that small-n settings create in general, how do we solve them? Through a combination of realigning our epistemology, using our current tools differently, and utilizing modern analytic tools developed outside the field of statistics, we can do better research and advance the field of space medicine to meet the challenges of the next 60 years.

3.1 Realigning our epistemology

Cognitive dissonance is the feeling of discomfort one feels when actions fail to conform to beliefs. [2] To most scientists, making claims about truth without a statistically significant result to point to elicits substantial cognitive dissonance. This perhaps more than anything demonstrates our over-reliance on NHST as a substitute for a more robust epistemology. There are several things we can do to learn from data without suffering from cognitive dissonance—even without significance tests. Altogether they amount to a different epistemological approach to epidemiology for space exploration.

3.1.1 Guidelines for causation

In 1965 Sir Austin Bradford Hill described nine guidelines for determining causation from scientific evidence. [3] It is worth noting that while one of the guidelines deals with *strength of association*, or what we might recognize as *effect size*, none of the criteria deal with significance testing or p-values. Explicitly, Hill called for examining the *quality* of the relationship between exposure and outcome: the logical features of how the evidence suggests they interact, and how that fits with prior knowledge of the same or similar subject matter. This sort of prescription is well-suited to the small-n environment of space medicine.

3.1.2 Modern causal inference theory: assumptions

Similar to Hill's work, modern causal inference methods may also be of great use in space-health research. These methods have sought to mathematically formalize causation in order to make valid use of observational data for causal estimation and to avoid introducing biases in analyzing such data [4]. Perhaps more important than the methods of analysis that this framework has promoted is the understanding of the assumptions necessary to make causal statements from non-randomized data. Merely understanding the assumptions of positivity, consistency, and conditional exchangeability—and what happens when one violates them—can be of tremendous help when trying to draw inferences based on limited data.

3.1.3 Directed acyclic graphs

A common tool used in modern causal inference is a special type of network graph known as the directed acyclic graph (DAG). These are network maps that reflect causal relationships. DAGs are drawn according to some simple rules, but making and using these diagrams can be quite useful for clarifying thinking and formulating testable hypotheses. If we factorize a joint probability distribution over a DAG, we create a Bayesian Network, a powerful tool of probabilistic inference. If we decompose a correlation or covariance matrix over a DAG, we can do path analysis or structural equation modeling, forms of latent-variable analysis. Even without any data collected at all, the structure of a DAG implies variable dependencies and independencies, which in turn have implications for what is and is not possible in the system from which the data were acquired, and thus can help guide critical thinking about problems.

3.1.4 Alternative hypotheses

A final epistemological realignment is to define specific, sensible hypotheses given the question at hand, which may or may not conform to the typical NHST two-tailed tests of significance. Examples of such alternatives include equivalence testing, inferiority testing, and a still more exotic choice, the *modus tollens*. All of these ask different questions than whether the central tendency of a sample shows enough difference to evince a significant p-value for the given sample size and variance. By changing the testable hypothesis to be more specific to what we really would like to know, we can often obtain an answer that is not only more sensible, but often more statistically powerful too, which might then bring NHST back into the realm of possibility to further refine the analysis.

3.2 Alternative analytic approaches

Yet another strategy for learning from data is the use of more-sophisticated analytic methods which do not necessarily rely on NHST. This includes exploiting properties of known statistical tests for alternative hypotheses, Bayesian methods, and machine learning.

3.2.1 Alternative uses of common statistical models

With a good understanding of common statistical models, it is possible to exploit their properties to conduct atypical investigations. Here we use an example from the literature on astronaut mortality to demonstrate this idea.

Using data on US astronauts and Soviet and Russian cosmonauts, Reynolds et al. [5] demonstrated that mortality from cancer and cardiovascular disease have no common causes in this population. This in turn was taken as evidence that doses of ionizing radiation received in space cannot have been sufficient to affect mortality from both of these causes. This was achieved by showing that a naïve analysis of survival curves (where competing causes of death were treated as censoring events) were not markedly different from survival curves that account for competing risks. That is, the causes of death displayed statistical independence which, in DAG terms, means they share no common ancestor.

In this example, the authors exploited the implications of different statistical methods for computing survival in presence of competing risks to make inferences regarding the structure of causal relationships. This is but one example, and undoubtedly others exist for those who can think broadly and conceptually about specific questions to be asked of existing datasets.

3.2.2 Simulation

The advancements in computing power over the last several decades have made possible more sophisticated forms of analysis, not least among them being simulation. We refer here to several different well-established approaches, all of which have found use in various domains such as statistics, business, and engineering.

Markov-chain Monte Carlo simulation (MCMC) has been used for decades in engineering for probabilistic risk assessment. Agent-based simulation has found increasing popularity in epidemiology for modeling community-level effects of policy change or change in social environment. Techniques such as the bootstrap and the jackknife may be loosely grouped here as well, as they rely upon repeated recalculation of sample statistics using algorithms that resample the data in specific ways. Finally, simple "what-if" analyses can help find the extremes of what is possible in a process or phenomenon, and can be used to eliminate possibilities or competing hypotheses.

3.2.3 Bayesian methods

Though certainly not new, Bayesian methods are still underutilized in research in general and in space medicine in particular. This is primarily owing to the unfamiliarity of most researchers with these methods, which in turn is due to the lack of graduate-level training on them in most scientific programs other than statistics. Historically, this was sensible: their mathematical complexity and need for computing power made them difficult to implement for all but the simplest of applications. Fortunately, computer science and computer hardware have both evolved to where these methods are easy to implement, creating a large opportunity for researchers to work with smaller datasets in meaningful and rigorous ways without reliance on NHST and p-values.

3.2.4 Data science

In recent years, Data Science has been turning business analytics upside down. In general, data science is understood as the science of learning from data, a seemingly perfect fit to our objectives here. Yet Data Science has seen much slower adoption in Academia, perhaps owing to the fact that the only part of Data Science that fits with the traditional epistemological approach to research is that part of Data Science which uses traditional NHST statistics.

A hallmark of Data Science is the use of machine learning. However, many of the methods of machine learning are methods that typically benefit from large datasets: those with hundreds of columns and millions of rows. Nevertheless, machine learning does have techniques that can be of use in the small-n world. Techniques for data reduction, data visualization, data mining, and simulation all are powerful tools that can often be applied in the domain of small-n research. Perhaps of particular interest to space medicine, researchers are able to use these methods for exploratory data analysis and hypothesis generation, tasks at which unsupervised machine learning excels.

4. Summary and conclusions

In this chapter we have discussed the limits of NHST as a surrogate for a broader, more flexible epistemological framework. Over-reliance on NHST can cripple the research enterprise when sample sizes and sampling schemes fail to conform to the assumptions necessary for valid models, much less valid inference.

A motivating factor for the use of NHST is the desire to draw correct conclusions. This is a valid aim, but may lead to an emphasis on error avoidance at the expense of learning from (possibly limited) data. Instead, scientists need to consider evidence using Hill's guidelines for causation, should examine whether or not the data in hand conform to or defy the assumptions needed for causal inference, and should include the use of DAGs to better understand what we already know about a given topic, and to clarify what we conjecture to be true *a priori*. Formulating so-called "alternative" hypotheses appropriate to the topic under study may even allow us to improve our inferences when using traditional NHST. There is no need to restrict ourselves to one approach or the other.

Alternative methods of analysis can be used to aid our understanding in small-data situations. Bayesian methods, more sophisticated uses of well-known statistical methods, and methods from data science all provide useful techniques that work well with small datasets, provided the scientist is willing to think differently about the outcomes of these analyses.

It is our hope that researchers involved in space medicine will adopt these perspectives and methods. To the extent that these ideas and techniques are adopted by the broader research community, we expect to see great advancements in our knowledge of health and safety in spaceflight. It is this expansion of our collective knowledge that will help make possible the space exploration missions of the next 60 years.

Author details

Robert J. Reynolds[1,2*] and Mark Shelhamer[3]

1 Mortality Research and Consulting, Inc., USA

2 Translational Research Institute for Space Health, Baylor College of Medicine, USA

3 The Johns Hopkins University School of Medicine, USA

*Address all correspondence to: rreynolds@mortalityresearch.com

References

[1] Ploutz-Snyder RJ, Fiedler J, Feiveson AH. Last word on viewpoint: Justifying small-n research in scientifically amazing settings: Challenging the notion that only "big-n" studies are worthwhile. Journal of Applied Physiology. 2014;**116**(9):1254. DOI: 10.1152/japplphysiol.00214.2014

[2] Festinger LA. Theory of Cognitive Dissonance. California: Stanford University Press; 1957

[3] Hill AB. The environment and disease: Association or causation? Proceedings of the Royal Society of Medicine. 1965;**58**(5):295-300. DOI: 10.1177/003591576505800503

[4] Pearl J. Causality. Cambridge: Cambridge University Press; 2009. DOI: 10.1017/CBO9780511803161

[5] Reynolds RJ, Bukhtiyarov IV, Tikhonova GI, Day SM, Ushakov IB, et al. Contrapositive logic suggests space radiation not having a strong impact on mortality of US astronauts and Soviet and Russian cosmonauts. Scientific Reports. 2019;**9**:8583. DOI: 10.1038/s41598-019-44858-0

Section 2

Changes in Physiology Due to Spaceflight

Chapter 2

Effects of Microgravity on Human Physiology

Satoshi Iwase, Naoki Nishimura, Kunihiko Tanaka and Tadaaki Mano

Abstract

The effects of microgravity conditions on neurovestibular, cardiovascular, musculoskeletal, bone metabolic, and hemato-immunological systems are described. We discuss "space motion sickness," sensorimotor coordination disorders, cardiovascular deconditioning, muscular atrophy, bone loss, and anemia/immunodeficiency, including their causes and mechanisms. In addition to the previously described deconditioning, new problems related to microgravity, spaceflight-associated neuro-ocular syndrome (SANS), and structural changes of the brain by magnetic resonance imaging (MRI) are also explained. Our proposed countermeasure, artificial gravity produced by a short-arm centrifuge with ergometric exercise, is also described in detail, and we confirmed this system to be effective in preventing the abovementioned deconditioning caused by microgravity exposure.

Keywords: microgravity, neurovestibular, neural plasticity, cardiovascular, musculoskeletal, bone metabolism, hematology and immunology

1. Introduction

Outer space offers several abnormal and/or unique environmental conditions, including microgravity, vacuum/hypovaria, acceleration, extreme temperature, space debris, space radiation, and confinement/isolation. As the latter four conditions may be mitigated by spacecraft engineering (i.e., pressurization and the bulkhead), we focused on microgravity and its effects on human physiology [1–8].

In spaceflight, astronauts face three periods of physiological adaptation induced by changing gravity: (1) changes upon entry to microgravity (initial adaptation), (2) changes after prolonged exposure to microgravity, and (3) readaptation to 1 G gravity on Earth after returning from space. Body systems influenced by microgravity are the neurovestibular, cardiovascular, musculoskeletal, bone metabolic, and immuno-hematological systems. The changes associated with these systems occur during the adaptation phases outlined above. We will briefly discuss each of these body systems.

2. Neurovestibular system

2.1 Space motion sickness

How do we humans sense our relative positions in three-dimensional space? There are three sensory systems in the human body that help us define position: the visual, somatosensory, and vestibular systems. Most of the information from the outside world is processed by the visual system, but the combination of somatosensory and vestibular systems from the inner body helps define the positional status.

The vestibular organs include the otolith organs and semicircular canals. The otolith organs, saccules (sagittal direction) and utriculi (horizontal direction), sense linear acceleration. The semicircular canals, anterior, posterior, and horizontal, detect angular velocity of the head. The vestibular organs in the inner ear detect and measure linear and angular acceleration. These responses—already a complex set of signals—are further integrated with visual and proprioceptive inputs.

Exposure to microgravity alters some of these input signals, leading to misinterpretation and inadequate responses by the brain. This may cause vertigo, nausea, vomiting, appetite loss, headache, pallor, etc. As the symptoms are like those of motion sickness, this set of symptoms is termed "space motion sickness," but unlike conventional motion sickness, antiemetic drugs cannot suppress the symptoms of space motion sickness. Approximately 60–80% of astronauts develop the symptoms within 2 or 3 days after launch. Space motion sickness is considered important because of its potential impact on the astronauts' operational performance.

Although sensory misinterpretation may play a role in space motion sickness, its exact mechanism remains unknown. There are, however, several hypotheses: (1) sensory conflict, (2) fluid shift, (3) otolith asymmetry, and (4) orientation adaptation [9, 10].

The sensory conflict hypothesis suggests that loss of tilt-related otolith upon entry into microgravity causes a conflict between actual and anticipated signals from sense organs subserving spatial orientation. Such sensory conflicts are thought to induce motion sickness in other environments.

The fluid shift hypothesis suggests that space motion sickness results from the caudad fluid shifts, which in turn result from the hydrostatic pressure gradients in the lower body to the thoracic cavity or to the cranial cavity when entering microgravity. The cephalad fluid shift leads to visible puffiness in the face and is thought to increase the intracranial pressure, the cerebrospinal fluid pressure, or the inner ear fluid pressure, altering the response properties of the vestibular receptors and inducing space motion sickness.

The otolith asymmetry hypothesis is based on the theory that a mass difference in otolith between the left and right ear is the origin of space sickness and that there is interindividual susceptibility.

The otolith adaptation theory, or otolith tilt-translation reinterpretation theory, is the theory that space motion sickness is caused during the process of the brain learning to reinterpret novel otolith quasi-static signals to represent linear acceleration in space rather than the usual interpretation of tilt relative to the vertical direction on Earth.

Until now, it was unclear which of these theories (if any) was most likely. However, evidence from Space Shuttle missions suggests that the otolith asymmetry and otolith adaptation theories are unlikely.

2.2 Countermeasures for space motion sickness

In the Shuttle program and in the case of the International Space Station (ISS), the most commonly used countermeasure for space motion sickness is pharmacotherapy. Dornhoffer [11] reported the effects of four drug countermeasures (lorazepam, meclizine, promethazine, and scopolamine) for alleviating motion sickness induced by vestibular stimulation with a rotary chair and found that scopolamine was the only countermeasure to significantly change the mean duration of rotation compared with the placebo ($p < 0.008$), with a > 40% increase in rotation time.

In the Shuttle study, administration of promethazine at 20–50 mg was recommended by intramuscular injection or suppository. In the ISS study, meclizine and dimenhydrinate with cinnarizine were hypothesized to affect the medial vestibular nucleus. Promethazine is a vestibular suppressor, but a more recent report [12] demonstrated that *d*-amphetamine counters this suppression and inhibits the effects of fatigue on the saccadic reaction time.

We propose that artificial gravity is also effective in preventing space motion sickness because constant gravity on the otolith is effective against all four etiologies of this maladaptation.

3. Cardiovascular system

3.1 Effects of microgravity in the cardiovascular system

The changes in the cardiovascular system begin solely with the fluid shift associated with microgravity, followed by the decreased circulatory blood volume, cardiac size, and aerobic capacity, and the most prominent symptom, postflight orthostatic intolerance. These symptoms are generically known as "cardiovascular deconditioning" [13–17].

When the spacecraft reaches low Earth orbit (LEO), body fluids move from the lower body to the thorax, which is associated with the increase in the intraocular pressure and morphological alterations in the central nervous system, demonstrated by changes in the magnetic resonance imaging (MRI).

As a result of fluid shift, the leg volume decreases and the face becomes puffy. The leg volume decreases by 1 L, whereas subcutaneous tissue at the forehead thickens by as much as 7% compared with in the preflight supine position. The pulmonary capillary blood volume increases by approximately 25%, and intraocular pressure can nearly double. Fluid shift increases the cardiac volume and stroke volume at the beginning of the spaceflight (first 24 to 48 hours), but over time, the heart rate, stroke volume, and cardiac output stabilize to the preflight sitting level. The arterial blood pressure slightly decreases compared with the preflight level. Compared with "space motion sickness," cardiovascular and fluid balance adaptation is gradual. The symptoms appear in 3–5 days and disappear after 1–2 weeks, causing facial edema, nasal stiffness, heavy headedness, papilledema, or jugular vein dilatation. These symptoms upon exposure to microgravity disappear at most 2 weeks after the reduction in circulatory plasma volume [13–17].

The cardiovascular changes in actual spaceflight differ from those in stimulations such as head-down bedrest or dry immersion. First, the volume of fluid shift is much larger than the orthostatic change from the supine to upright positions. The fluid volume loss during simulated microgravity (e.g., head-down bedrest or dry immersion) is less than 50% of that observed in actual spaceflight. Second, the central venous pressure measured during spaceflight does not increase as much as in head-down bedrest. Third, the diuresis caused during simulated microgravity is to a lower degree.

Then what is the cause of reduced blood volume after adaptation to microgravity? Diedrich et al. [18] explained the reduced blood volume in space as (1) a negative balance 2010 of decreased fluid intake and smaller reduction of urine output; (2) fast fluid shifts from the intravascular to interstitial spaces as a result of lower transmural pressure after reduced compression of all tissues by gravitational forces, especially of the thorax cage; and (3) fluid shifts from intravascular to muscle interstitial spaces because of lower muscle tone required to maintain body posture, and the attenuated diuresis during space flight is due to increased retention after stress-mediated sympathetic activation during the initial phase of space flight.

3.2 Decrease in the circulatory blood volume

The centralization of body fluid induces dehydration to adapt to the microgravity environment. The cephalad fluid shift causes an increase in venous return and marked increase in the stroke volume, inducing the alterations in the autonomic and endocrine systems to control the cardiovascular system.

On the first day of microgravity exposure, urine volume does not increase, but the circulatory blood volume suddenly decreases by 17%, probably due to the shift of water from the intravascular to the interstitial spaces and finally to the intracellular space. This induces an increase in the hematocrit level, which suppresses erythropoietin production and reduces the erythrocyte volume. Reductions in the circulatory plasma volume and erythrocyte volume equal an 11% reduction in the total blood volume, and this stabilizes the central blood volume to a new equilibrium, which nearly equals the central blood volume in the standing position at 1 G on Earth.

Upon the above alterations, the autonomic nervous system stabilizes the blood pressure by suppressing the sympathetic functions and activating the vagal functions by reducing the heart rate and suppressing muscle sympathetic nerve activity. Alterations include suppression of vasopressin by the Henry-Gauer reflex, facilitation of α-natriuretic peptide secretion, and suppression of the renin-angiotensin-aldosterone system, all of which facilitate urination. Thus, centralized body fluid is excreted, accounting for 10–15% of the circulatory blood volume, increased hematocrit level, and adaptation of the cardiovascular system 5–7 days after microgravity exposure. This ameliorates the facial edema and jugular distension. This adaptation causes cardiovascular deconditioning, including orthostatic intolerance, after returning to 1 G on Earth.

3.3 Reduced heart size

Once exposed to microgravity, the volume and pressure stimuli disappear. Constant postural change of lying down from upright standing on Earth loads intermittent volume on the heart, which ceases during microgravity. In addition, microgravity reduces the overall pressure load on the heart depending on the content of the countermeasure program. The mean arterial pressure slightly decreases. During spaceflight, the myocardial volume decreases by 8–10%.

3.4 Cardiovascular system after stabilization

After adaptation to microgravity, the cardiovascular system stabilizes, and the blood pressure is either unchanged or slightly lower [19]. Ambulatory blood pressure recording for 24 hours in eight astronauts revealed that the systolic, diastolic, and mean arterial pressures (mean ± se) in space were reduced by 8 ± 2 mmHg ($p = 0.01$; ANOVA), 9 ± 2 mmHg ($P < 0.001$), and 10 ± 3 mmHg ($p = 0.006$), respectively, with a maintained nocturnal dip of 8 ± 3 mmHg ($p = 0.015$). The cardiac stroke volume and output increased by 35 ± 10% and 41 ± 9% ($p < 0.001$), respectively, the heart rate and catecholamine concentrations were unchanged, and systemic vascular resistance was reduced by 39 ± 4% ($p < 0.001$).

3.5 Alteration of aerobic exercise capacity

Microgravity exposure reduces the circulatory blood volume; however, the maximal oxygen uptake is maintained after a short duration of spaceflight. During long-term spaceflight, the aerobic capacity decreases without countermeasures, but aerobic exercise training can maintain it, although standard exercise only markedly reduces the postflight maximal oxygen uptake. After a short duration of spaceflight (9–14 days), the maximal oxygen uptake decreased by 22%, probably due to decreases in the maximal stroke volume and maximal cardiac output without alterations in the maximal heart rate, blood pressure, or whole-body arteriovenous oxygen. This decrease in maximal oxygen uptake is believed to be due to the decreases in intravascular blood volume, stroke volume, and cardiac output.

As crew members are expected to work on the Moon/Martian surface, and they are exposed to extensive heat stress in the extravehicular suits, this aerobic capacity is considered to be significant after landing on the Moon/Mars.

3.6 Alterations in sympathetic neural traffic under microgravity

Sympathetic neural traffic indirectly measured by the plasma noradrenaline level has been reported to increase during spaceflight from the preflight control level [14, 20], and vagal activity estimated by power spectral analysis of heart rate variability was reduced after long-term spaceflight [21, 22].

Microneurographically recorded neural traffic in humans is known to be muscle and skin sympathetic nerve activity (MSNA and SSNA), and MSNA controls the vasomotor function of the muscular bed, responding to blood pressure changes against gravitational stress [23–25]. MSNA was suppressed during exposure to short-term microgravity induced by parabolic flight [26], mild lower body positive pressure (10–20-mmHg LBPP) [15], and thermoneutral head-out water immersion [27] responding to the loading or unloading of cardiopulmonary receptor-stimulated cephalad fluid shift. On the other hand, MSNA was increased after exposure to long-term microgravity in spaceflight and its simulation induced by dry immersion [28] or 6° head-down tilt bedrest [17] due to different mechanisms, including plasma volume loss, changes in baroreflex, and vascular compliance.

3.7 Postflight orthostatic intolerance

Orthostatic intolerance is usually observed after returning to 1 G on Earth. The definition of orthostatic intolerance usually includes simple syncope, lightheadedness, or >20-mmHg reduction in systolic blood pressure.

Astronauts usually notice orthostatic intolerance during prolonged upright standing rather than while standing up. Just before fainting, they sometimes have tachycardia, suggesting that they have postural orthostatic tachycardia syndrome (POTS). This phenomenon is due to a state in which fluid shift easily triggers tachycardia, which also easily triggers Bezold-Jarisch reflex, and the vagal response suppresses the systolic blood pressure. Although all astronauts stood upright for 10 min, 63% were unable to finish the stand test in 10 min.

Less important factors for postflight orthostatic intolerance are reduced compliance of the lower legs, reduced baroreflex sensitivity, and increased basal sympathetic tone.

Reduction of the circulatory blood volume is the most important factor for postflight orthostatic intolerance. The decrease in stroke volume after spaceflight reflects this circulatory blood volume loss. Although this is the main cause, the recovery of circulatory blood volume to the normal state is not complete. The crew members are recommended to take 8 g of salt and 1 L of water, which ameliorates the orthostatic tolerance, albeit not completely.

Another factor is the limitation by vasoconstriction. The postflight blood pressure of non-finishers cannot be increased by the total peripheral resistance compared with the preflight state. During the postflight upright standing 70° tilt test, the total peripheral resistance cannot increase despite activation of muscle sympathetic nerve activity, probably due to the alterations in venoarterial reflex and smooth muscle atrophy of the resistant vessels. Overall, circulatory blood volume reduction and attenuated vasoconstriction are the main factors for orthostatic intolerance.

3.8 Cardiovascular deconditioning

What is the cause of this cardiovascular deconditioning? NASA's criteria of orthostatic intolerance are (1) presyncopal symptoms (pallor, cold sweat, nausea, blackout, and fainting), (2) gradual systolic blood pressure decrease <80 mmHg, (3) sudden systolic blood pressure decrease >15 mmHg, or (4) sudden heart rate decrease >15 bpm while on the 70° tilt bed for 15 min. A recent report stated that 65% of astronauts satisfied these criteria. Previously, this cardiovascular deconditioning was considered to be solely due to circulatory fluid loss, but other causes have also been explored.

In addition to the decrease in circulatory blood volume, other causes, i.e., altered arterial baroreflex gain, altered leg venous volume, easy fluid pooling in the space of atrophied skeletal muscles, attenuated muscle pump effects due to skeletal muscle atrophy, hypersensitivity of β-adrenergic receptors, and altered influence of vestibular (especially otolith) input, have been considered. Moreover, increased venous permeability of lower leg vessels and attenuated cardiopulmonary volume receptor reflex after −6° head-down tilt for 14 days were observed in our bedrest experiment. These changes are not the only cause of cardiovascular deconditioning, and multiple factors act in concert.

3.9 Spaceflight-associated neuro-ocular syndrome (SANS)

Several physiological and pathological neuro-ocular findings in astronauts/cosmonauts during and after long-term spaceflight, including hyperoptic shifts up to +1.75 diopters, optic disc edema (swelling), globe (eyeball) flattening, choroidal folds, and "cotton wool" spots in the fundus oculi, have been reported [29]. These findings have been documented as spaceflight-associated neuro-ocular syndrome. NASA has investigated the clinical, ultrasound, optical coherence tomography imaging, and fundus oculi findings of the above symptoms. In 2016, out of 47 or 64 astronauts examined, approximately 10 developed SANS (disc edema in 10/64, cotton wool spot in 7/64, choroidal folds in 11/47, globe flattening in 12/47, and refractive error in 9/47). It is unlikely that the duration of spaceflight is unrelated [30].

The exact cause of SANS has not been clarified, but its development is likely related to the increase in intracranial pressure due to the cephalad fluid shift. The increase in intracranial pressure is not necessarily due to microgravity exposure, but some percentage of astronauts had intracranial pressure change and developed SANS [29, 30].

Several countermeasures, e.g., lower body negative pressure, thigh cuffs, an impedance threshold device (ITD), vitamin B group administration, and artificial gravity, have been considered and are under trial. NASA and collaborating researchers continue to investigate SANS in preparation for future manned missions to space, including continued trips to the ISS, deep space gateway missions, a return to the Moon or Moon base, or a Martian expedition.

3.10 Brain structural plasticity during spaceflight

In 2016, structural changes in the brain during spaceflight were reported. Koppelmans et al. [31] evaluated retrospective longitudinal T2-weighted MRI scans and balance data from 27 astronauts (13, ~2-week Shuttle crew members, and 14, ~6-month ISS crew members) to assess spaceflight effects on brain structure. They observed extensive volumetric gray matter decreases, including large areas covering the temporal and frontal poles and around the orbits, and the effects were larger in ISS members than in Shuttle crew members. There were also bilateral focal gray matter increases within the medial primary somatosensory and motor cortex.

In 2017, a review on these MRI changes associated with spaceflight (actual or simulated) was reported. Van Ombergen et al. [32] discussed neuroplastic changes in the central nervous system and concluded that the cerebellum, cortical motor areas, and vestibular-related pathways are highly involved, demonstrating that these brain regions are indeed affected by actual and simulated spaceflight. Structural studies are now in progress, and functional relationships are under investigation. Long-term studies will be necessary to clarify the mechanism.

3.11 Effects of artificial gravity

We tested an intermittent short-arm centrifuge of 1.4 G with 60-W ergometric exercise with a step-up increase of 0.2 G and 15 W, respectively, for 30 min every day for 21 days during −6° head-down bedrest [33]. The circulatory blood volume was reduced by 20% in the control subjects, but no reduction was observed in the countermeasure subjects. Cardiac output and stroke volume were not changed in the countermeasure subjects, but they decreased in the control subjects. The baseline level of muscle sympathetic nerve activity (MSNA) was not changed in the countermeasure subjects, but it increased in the control subjects.

Therefore, everyday ergometric exercise under artificial gravity maintains the preflight cardiovascular state without adapting to microgravity.

4. Musculoskeletal system

4.1 Mechanism of muscle loss under microgravity

The first muscular measurements were performed in Skylab and Space Shuttle missions by the United States and in Salyut and Mir by the Soviet Union [34–37]. The most prominent muscle loss was observed in the calf muscle (the soleus and gastrocnemius) after a few weeks in space. The muscle loss exhibited interindividual variation, but the maximum loss reached as high as 10%. This volume loss in the lower extremities accounts for most of the muscle atrophy and the blood and interstitial fluid shift. Although fluid shift away from the legs influences the size of these muscles, this phenomenon alone cannot explain the changes in leg volume on MRI. Muscle atrophy appears rapidly, usually between 8 and 11 days of flight, but can appear as early as the fifth day, as observed in one astronaut. Moreover, the effects of microgravity differ among muscles, with volume decreasing by 3.9% in the calf (the soleus and gastrocnemius) and 6% in the quadriceps femoris.

In addition to the morphological changes, functional alterations are associated with structural variations, and the muscular force is known to be reduced after spaceflight. In the Skylab 3 mission, it decreased by 20% after 53 days of microgravity exposure. Muscular electrical activity measured by electromyogram (EMG) had a lower EMG amplitude in addition to easy fatigability with a lower resistance.

The main cause of muscular loss is the disappearance of mechanical constraints and the subsequent decrease in muscular activity. The reduced muscular activity under microgravity is also associated with hypokinesia due to limited movement inside the spacecraft, which also can be observed in bedrest studies and animal experiments using tail suspension. The structural changes in skeletal muscle are also observed by microscopic examination under microgravity, which revealed that the proportion of type I red fibers decreased and they were replaced by type II white fibers. In 1995, pre- and postflight human muscle biopsies were performed on three and five astronauts during 5- and 11-day missions, respectively. In this study, the muscle fiber diameter decreased by 15 to 30%, and the number of capillaries around the muscle fibers decreased. The proportion of type I fibers changed from 43 (preflight) to 37% (postflight) after 5-day missions, and that of type II fibers changed from 57 to 67%. After 11 days of spaceflight, the proportion of type I fibers decreased from 45 to 39%, and that of type II fibers decreased from 55 to 61%, consistent with the animal experiments that demonstrated that gravity can influence genes regulating the protein synthesis of muscle protein degradation enzymes.

These changes were confirmed to be due to both an increase in protein breakdown and decrease in synthesis. Human biochemical examination also revealed a higher level of muscle protein degradation, increased level of urinary amino acids, and higher level of creatinine. The diet of astronauts is protein-rich, but the degradation process is such that nitrogen losses overcome the gains and the nitrogen balance becomes negative.

Recent studies have suggested the mechanism of disuse atrophy of the skeletal muscle, especially oxidative stress, to be an important regulator of pathways leading to muscle atrophy during periods of disuse. Redox disturbances, such as those in skeletal muscle myotubes, increase the expression of key components of the proteasome proteolytic system, which is a prominent factor in protein degradation in disused muscles.

Another hypothesized mechanism is the degradation of muscle proteins resulting from their ubiquitination. These molecular mechanisms underlie protein degradation during disuse.

4.2 Countermeasures for muscle loss

In order to prevent muscle loss, several countermeasures were available during microgravity exposure in Shuttle missions and continue to be available on the ISS. These include aerobic exercise, stretching, strength training, and electrical stimulation. Artificial gravity with exercise has been proposed as a potential measure for muscle loss because ground-based studies confirmed it to be effective, but a short-arm centrifuge has not been mounted. In addition to physical stimulation, medications, such as antioxidants, growth hormone, growth factors, ubiquitin, clenbuterol, anabolic steroids, and amino acids, are candidates against muscle loss.

Aerobic exercise is the most effective countermeasure for maintaining the fast twitch red fibers. On the ISS, the combination of an ergometer, treadmill, and the Advanced Resistive Exercise Device (ARED) is ideal to maintain muscle power and the morphology of antigravity muscles (empowerment of both fast and slow twitch fiber muscles).

Stretching can minimize atrophy by maintaining the muscle as much as possible in the stretched condition. In the Russian space program, the penguin suit—a snug fitting, full-length, long sleeved jumpsuit made with elastic inserts at the collar, waist, wrists, and ankles and along the vertical sides of the suit—loads the body along the long axis with an adjustable force of 15–40 kg, while the other elastic elements make it possible to adjust the position of the limbs. The angle of the major joints, such as the knee and ankle, can be set, allowing the foot to be dorsiflexed, which will stretch the soleus. The effects of stretching are not well understood, but it is a valid countermeasure considering human physiology.

Strength training by resistance exercise is also employed as a countermeasure for muscle loss. Weight training studies recommended a good mix of exercise types to be 15% (eccentric), 10% (isometric), and 75% (concentric weight-bearing). The most recommended exercise for resistance training is squatting.

Electric stimulation is also expected to increase protein synthesis and prevent the decrease in oxidative enzymes inducing disuse atrophy.

4.3 Artificial gravity with exercise

We previously confirmed the effectiveness of artificial gravity in bedrest studies [33]. Our results demonstrated that artificial gravity of 1.4 G with ergometric exercise maintains the muscle strength and cross-sectional area of the quadriceps femoris measured by MRI. Another study revealed that artificial gravity with squatting exercise also maintains the function and morphology of the soleus and gastrocnemius [38].

5. Bone metabolism system

The main problem of the skeletal system is bone calcium (Ca^{2+}) loss during microgravity. The bone becomes fragile during microgravity exposure, which can harm an astronaut or cosmonaut even after returning to Earth. Moreover, the risk of renal stones is high during long-term missions due to hypercalcemia [39].

Ca^{2+} plays an essential role in bone structure, contraction of skeletal and cardiac muscles, neural transmission, blood coagulation, cell permeability, and hormonal signaling. The serum Ca^{2+} level is well maintained at 8.4–10.2 mg/dL. Ca^{2+} is absorbed from the small intestine (300 mg/day), into the blood, deposited in the bone (500 mg/day), and excreted from the kidneys (150 mg/day) or into feces.

In this section, the influences of gravity on bone structure and of hormones on bone formation and absorption are described.

5.1 Bone development and restructuring

Gravity influences the long bones of the lower extremities, e.g., the femur, tibia, calcaneus, and vertebrae, which support the body in the upright position. Bone tissue contains osteocytes, which develop from osteoblasts, and are changed into osteoclasts by the action of RANKL (also called osteoclast differentiation factor, short for receptor activation of NF-kappa B ligand).

Osteoblasts and osteoclasts are functionally closely related, as is the balance between bone formation and bone resorption. Thus, insufficient bone formation compared with bone resorption observed in spaceflight reduces the bone mass and bone strength, leading to fractures.

Two hormones, calcitonin and parathormone (PTH), and vitamin D play an essential role in Ca^{2+} metabolism. Calcitonin is secreted from C cells of the thyroid gland. The secretion of calcitonin is promoted by an increase in the blood calcium concentration and is suppressed by the decrease in the serum calcium level. Calcitonin acts on the calcitonin receptor in osteoclasts to suppress the release of Ca^{2+} from the bone and promotes the deposition of calcium and phosphate on the bone. Calcitonin also promotes the excretion of calcium and phosphate into the urine. As a result, the serum level of Ca^{2+} decreases.

PTH is secreted from the parathyroid gland, increasing the release of Ca^{2+} from bone. PTH binds to osteoblasts to increase the expression of RANKL and inhibits their secretion of osteoprotegerin (OPG), which competitively binds to RANKL, preventing RANKL from interacting with RANK (receptor for RANKL). The binding of RANKL stimulates osteoclast precursors to fuse, forming new osteoclasts. As a result, PTH increases bone resorption, thereby increasing the serum Ca^{2+} level.

Vitamin D (25-hydroxycholecalciferol) acts on the parathyroid gland and suppresses the synthesis and secretion of PTH. The intestinal tract promotes the absorption of calcium and phosphorus. Vitamin D is essential for bone formation, but its direct action on bone formation remains unclear. It is also necessary for the formation of osteoclasts, affects the bone action of PTH, and promotes bone resorption itself at a high concentration. In the kidney, vitamin D increases Ca^{2+} reabsorption in the distal tubule and promotes the Ca^{2+} reabsorption action of PTH.

5.2 Effects of microgravity on bone metabolism

The main factor of bone metabolism is mechanical impact. Gravity creates weight and is responsible for the pressure exerted on a large part of the skeleton, resulting in a mechanical impact on bones. These gravitational impacts provide mechanical constraints on the femurs, tibias, calcaneus, and vertebrae. Thus, decalcification and bone loss are observed as the result of bone resorption, and the disappearance of gravity from the body axis components induces bone loss and resultant osteoporosis [2, 40].

5.3 Effects of spaceflight on bone metabolism

In microgravity conditions, decalcification was observed in 12 of the astronauts on the Gemini and Apollo 7 and 8 flights in 1969 [8]. Based on bone density measured using X-rays, the bone Ca^{2+} loss was 3.2%.

The bone mineral density was measured before and after the Mir program. The changes were as follows: +0.6% in the skull, +0.1% in the arm, −1.07% in the spine, −1.35% in the pelvis, −1.16% in the femoral neck, −1.58% in the trochanter major, −1.25% in the tibia, and −1.50% in the calcaneus per month, with comparable results from the ISS [41]. The most affected bones during spaceflight are weight-bearing bones, e.g., the pelvis (os coxae), the trochanter major of the femur, the femoral neck, the tibia, and the calcaneus.

Mir studies using dual photonic densitometry demonstrated a mean decrease in bone mineral density of approximately 0.3%/month in the cortical bone and up to 0.9%/month in cancellous tibial bone. Decalcification only occurs in the weight-bearing bones, and demineralization is correlated with mission duration. During spaceflight, hypercalcemia and hyperphosphatemia develop due to demineralization, and Ca^{2+} excretion into the urine increases before stabilization at around the 30th day of flight [42].

The National Aeronautics and Space Administration (NASA) of the United States documented a bone mineral loss rate/month of 1–2% during spaceflight (https://science.nasa.gov/science-news/science-at-nasa/2001/ast01oct_1), and bone density loss of 5–6% was reported in Apollo 15 crewmembers [43]. Bone loss is influenced by the spaceflight duration.

Tail suspension studies on rats demonstrated that simulated microgravity reduces bone formation, alters the Ca^{2+} balance, and inhibits the proliferation and differentiation of osteoprogenitor cells [44]. Osteocytes can also be affected by unloading stimulus in a bioreactor, with high expression of inhibitors of bone formation (sclerotin) and stimulators of bone resorption (RANKL) [45].

Osteoporosis can be irreversible. After returning to Earth without appropriate rehabilitation, the bone may be unable to return to normal activity under loads and may be weaker, easily inducing fractures. Reloading after a period of a week can ameliorate bone weakness, but even 2 weeks of bone restoration was not satisfactory. The length of spaceflight influences the bone density loss; however, another report stated that the bone loss at 1 month continued to increase after 6 months, suggesting that the length of spaceflight does not determine the bone density loss [42].

5.4 Effects of mechanical impact on hormonal influence

Wolff's law characterizes how the bone adapts to functionally withstand its mechanical environment [46]; however, several studies found that mechanical loading per se is not the direct stimulus for bone remodeling [47–51].

Then what are the effects of mechanical impact on calcitonin? Calcitonin receptors were not observed on osteoblasts [52] but were present on osteoclasts [53] and osteocytes [54]. Calcitonin was reported to inhibit apoptosis of osteoblasts and osteocytes, demonstrating a potential indirect influence on bone formation. These data confirmed that mechanical impact is not directly related to bone formation.

One notable property of PTH is that although chronic increases in PTH levels increase bone resorption, intermittent stimulation accelerates bone formation. PTH stimulates osteoclast formation by binding to PTH receptor 1 on stromal/osteoblastic cells and thereby increases the production of receptor activator of RANKL and macrophage colony-stimulating factor (M-CSF) and suppresses the RANKL decoy receptor osteoprotegerin. Moreover, PTH controls the production of osteoblasts through actions on osteocytes through Wnt signaling in osteoblastogenesis [55].

The action of osteocytes, which can directly sense a mechanical unloading stimulus, increased the expression of both inhibitors of bone formation (SOST/sclerotin) and stimulators of bone resorption (RANKL) through Wnt signaling [45]. These results support the hypothesis that intermittent mechanical impacts induce osteocyte action, which inhibits bone formation and stimulates bone resorption, and that an intermittent increase in PTH controls the production of osteoblasts.

5.5 Sympathetic alteration of bone metabolism during spaceflight

It has been reported that sympathetic neural traffic to bone inhibits the function of osteoblast and increases that of osteoclast, thus facilitating bone loss. Possible roles of the sympathetic nervous system in the mechanisms of bone loss in humans exposed to long-term spaceflight will be discussed.

Prolonged exposure to microgravity in space for 14 days increased sympathetic neural traffic in humans based on results from the Neurolab mission [14, 56–58], with comparable increases in noradrenaline spillover and clearance in space [14]. Concordant results were obtained during simulated microgravity, including dry immersion [28] or head-down bedrest [17]. In general, elderly people have a low bone density and high sympathetic neural traffic to muscles [59]. Our preliminary study demonstrated that changes in sympathetic neural traffic to muscles after long-term bedrest of 20 days were significantly correlated with changes in the urinary secretion level of deoxypyridinoline [25, 60], which is used as a specific marker of bone resorption [61]. Based on these findings, exposure to prolonged microgravity may increase sympathetic neural traffic to the bone, which increases the noradrenaline level, thereby inhibiting osteogenesis and facilitating osteolysis through β-receptors to induce bone mineral loss.

5.6 Countermeasures for space-related osteoporosis

1. Physical factors

Exercise during weightlessness has been incorporated into the present countermeasure programs; however, exercise alone cannot prevent bone loss. The current exercise program for the ISS is a combination of aerobic and resistive exercise for 2.5 hours, 6 days/week. Data from spaceflight revealed that bone loss occurs mainly in the femur, tibia, calcaneus, and vertebrae. Therefore, exercise should be concentrated on these bones, and impact loading should be primarily provided rather than static loading [62].

2. Pharmacological factors

As bone mass is sufficient at the onset of the spaceflight, the optimal strategy for the pharmacotherapy against bone loss is the prevention of bone loss, not the acceleration of bone formation, when loading is removed during spaceflight. Several drugs have been proposed to prevent bone loss under microgravity.

3. Bisphosphonates

Bisphosphonates have two phosphonate (PO_3) groups and are similar in structure to pyrophosphate. They bind to hydroxyapatite in bone matrix and prevent bone loss by inhibiting osteoclastic bone resorption. Bisphosphonates have been demonstrated to be effective in preventing bone loss during bedrest studies [63–66]. Among several types of bisphosphonates, pamidronate has been confirmed to suppress bone mineral loss and to prevent the formation of renal stones during bedrest studies [67].

In 2010, LeBlanc and Matsumoto [68] proposed an experiment for the effectiveness of bisphosphonate as a countermeasure to spaceflight-induced bone loss. The astronauts chose either oral administration of alendronate at 70 mg once per week or intravenous administration of zoledronate at 4 mg before the flight, and their bone densities were examined by DXA, QCT, and pQCT, and bone metabolism markers, including bone formation and resorption markers, and renal stone formation were assessed. One of the co-investigators (Ohshima) reported successful suppression of spaceflight-induced bone loss and renal stone formation (Ohshima, personal communication).

The disadvantages of bisphosphonates are local irritation of the upper gastrointestinal (GI) tract and poor absorption from the GI tract. Therefore, the oral administration of bisphosphonates requires drinking 200 mL of water while remaining in an upright posture for at least 30 min until after their first meal of the day to facilitate delivery to the stomach. This poses a problem as there is no upright posture in space due to microgravity. Another potential problem is osteonecrosis of the maxilla and the mandible, although the incidence is low [69]. These osteonecrotic or osteolytic phenomena always accompany physiological stress (mastication), iatrogenic trauma (tooth extraction/denture injury), or tooth infection [70, 71].

Bisphosphonates are difficult to metabolize, and high concentrations of them are maintained in the bones for long periods. As bone formation is closely coupled with bone turnover, long-term use of the compound with resultant suppression of bone turnover can compromise healing of even physiological microinjuries within the bone. Osteonecrosis of the maxilla and mandible likely results from the inability of the hypodynamic and hypovascular bone to meet the increased demand for repair and remodeling because several alterations are associated with this necrosis.

4. Parathormone

Parathormone has anabolic effects on the bone and also functions in the kidney to stimulate the reabsorption of $Ca2^+$ and increase the synthesis of vitamin D. In this sense, parathormone may stimulate bone formation, increase vitamin D synthesis, and stimulate $Ca2^+$ reabsorption. As suppressing bone reabsorption is favorable for stimulating bone formation during spaceflight, the administration of parathormone is strategically unfavorable.

In conclusion, it is favorable to administer bisphosphonate orally under artificial gravity with exercise in order to prevent osteoporosis in space. Monitoring the blood and urine samples on the ISS or spacecraft by a simple method is necessary to assess the effectiveness of the countermeasure.

6. Immunology and hematology

6.1 Space anemia

The circulatory blood volume is 5 L on average and contains plasma and cellular components, including erythrocytes, leukocytes (neutrophils, eosinophils, basophils, and lymphocytes), and platelets [72–74]. Among them, reduction in cellular components, especially erythrocytes (RBC), is associated with anemia, whereas the function of leukocytes is related to immunological response.

In the early stages of space development, cases of "space anemia" (hematocrit reduction) were reported on Gemini, Apollo, Skylab, and Shuttle missions and in the cosmonauts in the Salyut and Mir missions. However, in spaceflight, microgravity causes cephalad fluid shift, meaning this "space anemia" was actually a misinterpretation of symptoms. The true effects of microgravity can be measured through the total RBC count calculated from the hematocrit and plasma volume measurements. In this way, "space anemia" corresponded to a reduction in the total RBC count.

After 10 days aboard Spacelab-1, the total RBC count was reduced by 9% and by 15% after several weeks. After returning to Earth, the total RBC count did not recover even after 6 weeks and, in the case of Skylab astronauts, had not recovered after more than 3 months.

This suggested that microgravity is responsible for "space anemia," and many investigations were carried out to reveal whether anemia is the result of an increase in RBC destruction or a decrease in their production. Using labeled RBC by the uptake of ^{14}C glycine, RBC destruction was found to be three times greater in rats having flown aboard Cosmos-782 than in the control rats. On the other hand, reduced RBC production is unlikely because the number of stem cells measured by the number of cellular colonies that developed in vitro from samples of bone marrow taken from rats that flew aboard the Soviet Biosatellite—2044 for 14 days—was unchanged.

Human studies carried out by [75] on Shuttle missions for 9 to 14 days demonstrated that space anemia is due mainly to a lower production of RBC, causing increased plasma volume, reduced hemoglobin concentration, and increased serum erythropoietin. This reflects a decrease in the RBC life span and slower production.

Rizzo et al. [76] analyzed the cause of the shortened RBC life span and reported altered cell membrane composition and an increase in lipid peroxidation products. They suggested that antioxidant defense systems in the erythrocytes were induced, with a significant increase in glutathione content.

The mechanism underlying anemia was also confirmed by measuring the erythropoietin (EPO) level [77]. Radioimmunoassay revealed that the EPO level decreases after 24 hours of flight and is reduced by 30–40% on the third day compared with preflight levels. This low secretion of EPO will inhibit RBC maturation and cause hemolysis due to suppressed erythropoiesis.

Other changes in leukocytes (WBC) are in their polymorphonuclear characteristics. The composition of WBC is changed such that there is a slight increase in neutrophils and decrease in eosinophils. The percentage of lymphocytes, especially T cells, decreases, whereas that of monocytes slightly increases. These changes quickly disappear upon returning to 1 G on Earth.

6.2 Immunological changes during weightlessness

Recent studies confirmed dysregulation of the immunological response in humans and the reactivation of latent herpes virus, which persisted for the duration of a 6-month orbital spaceflight [78]. Blood samples from ISS crew members demonstrated that long exposure to microgravity reduced their T lymphocyte counts, suggesting the attenuation of cytotoxic function and viral reactivation in the space environment.

As the immune system is highly sensitive to different types of stressors, including psychological, physical, and local environmental stressors (e.g., oxidative and radiation exposure), exposure to the space environment suppresses T helper cells, which leads to susceptibility to viruses.

7. Artificial gravity as a total countermeasure for spaceflight deconditioning

For human space voyages lasting several years, such as those envisioned for the exploration of Mars, astronauts will be at risk of catastrophic consequences should any of the systems that provide air, water, food, or thermal protection fail. Beyond that, astronauts will face serious health and/or safety risks resulting from marked physiological deconditioning associated with prolonged weightlessness [1, 79]. The principal physiological deconditioning risks are related to physical and functional deterioration, and the loss of regulation of the several systems, including blood circulation, decreased aerobic capacity, musculoskeletal systems, and altered sensorimotor system performance. These physiological effects of weightlessness are generally adaptive to spaceflight and present a hazard only following G transitions upon returning to Earth or landing on another planet [80]. Among them, bone mineral metabolism will be greatly affected during prolonged spaceflight.

7.1 Why artificial gravity

Space biomedical researchers have been working for many years to develop "countermeasures" to reduce or eliminate the deconditioning associated with prolonged weightlessness. Intensive and sustained aerobic exercise on a treadmill, bicycle, or rowing machine coupled with intensive resistive exercise has been used on US and Russian spacecraft to minimize these problems. The procedures were uncomfortable and excessively time-consuming for many astronauts, and their effectiveness for maintaining bone, muscle, and aerobic fitness has not been demonstrated due in part to the low reliability of the devices flown to date. Furthermore, they have had inconsistent effects on postflight orthostatic hypotension or sensorimotor adaptive changes. With the exception of fluid loading before reentry, other countermeasures (e.g., diet, lower body negative pressure, or wearing a "penguin suit" to force joint extension against a resistive force) either have been marginally effective or presented an inconvenience or hazard.

To succeed in the near-term goal of a human mission to Mars during the second quarter of this century, the human risks associated with prolonged weightlessness must be mitigated well beyond our current capabilities. Indeed, during nearly 45 years of human spaceflight experience, including numerous long-duration missions, no single countermeasure or combination of countermeasures that is completely effective has been developed. Current operational countermeasures have not been rigorously validated and have not fully protected any long-duration (>3 months) astronauts in low Earth orbit. Thus, it is unlikely that they will sufficiently protect astronauts journeying to Mars and back over a 3-year period.

Although improvements in exercise protocols, changes in diet, or pharmaceutical treatments of individual systems may be of value, they are unlikely to eliminate the full range of physiological deconditioning. Therefore, a complete research and development program aimed at substituting the missing gravitational cues, and loading in space is warranted.

The urgency of exploration-class countermeasures is compounded by the limited availability of flight resources for validating a large number of system-specific countermeasure approaches. Furthermore, recent evidence of the rapid degradation of pharmaceuticals flown aboard long-duration missions, putatively because of radiation effects, raises concerns regarding the viability of some promising countermeasure development results. Although the rotation of a Mars-bound spacecraft will not be a panacea for all human risks of spaceflight (artificial gravity cannot solve the problems associated with radiation exposure, isolation, confinement, and environmental homeostasis), artificial gravity does offer significant promise as an effective, efficient, multi-system countermeasure against the physiological deconditioning associated with prolonged weightlessness. Virtually all of the identified risks associated with cardiovascular deconditioning, myatrophy, bone loss, neurovestibular disturbances, space anemia, immune compromise, and neurovegetative state may be alleviated by sufficient artificial gravity.

7.2 Why artificial gravity with exercise

Although a short-radius centrifuge has been proposed several times, loading with artificial gravity has not been demonstrated to be effective at preventing spaceflight deconditioning on its own. Making a human-powered short-arm centrifuge is an effective method to create exercise loads for astronauts. Considering the size of the ISS, it is appropriate to employ a short-radius centrifuge rather than a large-radius human centrifuge; however, it may be beneficial to rotate the spacecraft itself to provide the artificial gravity for long-duration spaceflight such as Mars expeditions.

In 1999, Iwase proposed the creation of artificial gravity by ergometric exercise, and it was installed at Nagoya University [33]. Several studies were performed using this short-radius centrifuge with an ergometer. In 2002, a bedrest study was carried out to evaluate the effectiveness of artificial gravity with ergometric exercise. In 2005, the facility was moved to Aichi Medical University, and bedrest studies

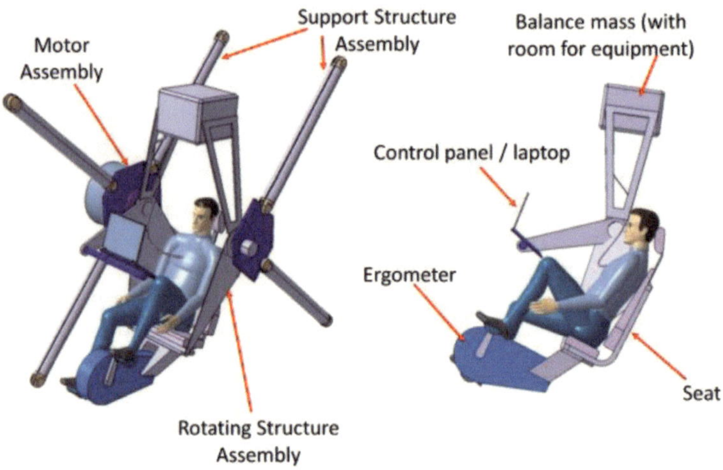

Figure 1.
Structure of the device of artificial gravity with exercise for AGREE project 2012.

were performed to finalize the protocol. This daily AG-EX step-up protocol was confirmed to be effective at preventing cardiovascular, musculoskeletal, and bone metabolism deconditioning in 2006, whereas the alternate-day (every-other-day) protocol (loading the AG-EX every other day) did not improve the spaceflight deconditioning associated with the microgravity exposure analogue of −6° head-down bedrest.

The authors applied for the installation of a short-radius centrifuge facility on the ISS and proposed it as a method to prevent spaceflight deconditioning, including bone loss. This project, Artificial Gravity with Ergometric Exercise (AGREE project), was promising to prevent space deconditioning during spaceflight, but it was canceled halfway through (**Figure 1**).

8. Conclusion and summary

Several deconditioning states in the neurovestibular, cardiovascular, ocular, musculoskeletal, bone metabolic, hematological and immunological, and central nervous systems have been documented, and efforts to ameliorate the symptoms have been made. In the near future, space medicine will play an increasingly important role in missions to the Moon and Martian expeditions as well as in future deep space exploration.

Historically, space medicine examined early adaptation to microgravity and early readaptation to the terrestrial 1 G state. However, the philosophy of the authors is to avoid adaptation to microgravity using artificial gravity. Under this scenario, short exposure to microgravity is permitted, but longer adaptation will be unnecessary. With this philosophy, the authors believe that the humans will achieve safe and comfortable spaceflight without deconditioning.

Author details

Satoshi Iwase[1*], Naoki Nishimura[2], Kunihiko Tanaka[3] and Tadaaki Mano[3]

1 Department of Physiology, Aichi Medical University, Japan

2 Nihon Fukushi University, Japan

3 Gifu University of Medical Sciences, Japan

*Address all correspondence to: s_iwase@nifty.com

References

[1] Buckey JC Jr. Preparing for Mars: The physiologic and medical challenges. European Journal of Medical Research. 1999;**4**:353-356

[2] Buckey JC Jr. Space Physiology. New York: Oxford University Press; 2006. pp. 1-283

[3] Clément G. Fundamentals of Space Medicine. 2nd ed. New York: Springer; 2011. pp. 1-381

[4] International Academy of Astronautics Study Group, editor. Bone standard measures. In: International Academy of Astronautics Study Group. In: Guidelines for Standardization of Bed Rest Studies in the Spaceflight Context. 2012a. pp. 44-46

[5] International Academy of Astronautics Study Group. Guidelines for standardization of bed rest studies in the spaceflight context; 2012b

[6] Linenger J. Off the Planet: Surviving Five Perilous Months Aboard the Space Station MIR. New York, NY: MacGraw-Hill; 1999. pp. 1-272

[7] Phillips RW. Grappling with Gravity. How Will Life Adapt to Living in Space? New York, Dordrecht, Heidelberg, London: Springer; 2012. pp. 1-271

[8] Planel H. Space and Life, an Introduction to Space Biology and Medicine. Boca Raron, London, New York, Washington, D.C.: CRC Press; 2004. pp. 1-178

[9] Koizuka I, Kaato I. Space motion sickness and spatial orientation of vestibulo-ocular reflex. Equilibrium Research. 1999;**58**:9-20

[10] Scherer H, Helling K, Clarke AH, Hausmann S. Motion sickness and otolith asymmetry. Biological Sciences in Space. 2001;**15**:401-404

[11] Dornhoffer JL. Pharmacological Countermeasures for Space Motion Sickness. Report from National Space Biomedical Institute. 2004. Available from: http://nsbri.org/researches/pharmacological-countermeasures-for-space-motion-sickness/ [Accessed: 27 September 2019]

[12] Weerts AP, Vanspauwen R, Fransen E, Jorens PG, Van de Heyning PH, Wuyts FL. Space motion sickness countermeasures: A pharmacological double-blind, placebo-controlled study. Aviation, Space, and Environmental Medicine. 2014;**85**:638-644

[13] Antonutto G, di Prampero PE. Cardiovascular deconditioning in microgravity: Some possible countermeasures. European Journal of Applied Physiology. 2003;**90**:283-291

[14] Ertl AC, Diedrich A, Biaggioni I, Levine BD, Robertson RM, Cox JF, et al. Human muscle sympathetic nerve activity and plasma noradrenaline kinetics in space. The Journal of Physiology. 2002;**538**:321-329

[15] Fu Q, Sugiyama Y, Kamiya A, Shamsuzzaman ASM, Mano T. Responses of muscle sympathetic nerve activity to lower body positive pressure. The American Journal of Physiology. 1998;**275**:H1254-H1259

[16] Grigoriev AI, Morukov BV, Vorobiev DV. Water and electrolyte studies during long-term missions onboard the space stations SALYUT and MIR. The Clinical Investigator. 1994;**72**:169-189

[17] Kamiya A, Iwase S, Kitazawa H, Mano T, Vinogradova OL, Kharchenko IB. Baroreflex control of muscle sympathetic nerve activity after 120 days of 6 degrees head-down bed rest. American Journal of Physiology.

Regulatory, Integrative and Comparative Physiology. 2000;**278**:R445-R452

[18] Diedrich A, Paranjape SY, Robertson D. Plasma and blood volume in space. The American Journal of the Medical Sciences. 2007;**334**:80-85

[19] Norsk P, Asmar A, Damgaard M, Christensen NJ. Fluid shifts, vasodilatation and ambulatory blood pressure reduction during long duration spaceflight. The Journal of Physiology. 2015;**593**:573-584

[20] Christensen NJ, Norsk P. Sympathoadrenal activity is increased in humans during spaceflight. Journal of Gravitational Physiology. 1998;5:P13-P14

[21] Cooke WH, Ames JE IV, Crossman AA, Cox JF, Kuusela TA, Tahvanaine KUO, et al. Nine months in space: Effects on human autonomic cardiovascular regulation. Journal of Applied Physiology. 2000;**89**:1039-1045

[22] Mano T. Autonomic neural functions in space. Current Pharmaceutical Biotechnology. 2005;**6**:319-324

[23] Iwase S, Mano T, Saito M. Effects of graded head-up tilting on muscle sympathetic nerve activities in man. Physiologist. 1987;**30**:S62-S63

[24] Mano T. Sympathetic nerve mechanisms of human adaptation to environments - findings obtained by recent microneurographic studies. Environmental Medicine. 1990;**34**:1-35

[25] Mano T, Iwase S, Nishimura N, Fu Q, Cui J, Shamsuzzaman AS, et al. Gravitational stress on the sympathetic nervous system in humans. In: Invasive and Non-invasive Studies of the Human Autonomic Nervous System. A Satellite Meeting of ISAN2009. Sydney; 2009. pp. 22-23

[26] Iwase S, Mano T, Cui J, Kitazawa A, Kamiya A, Miyazaki S, et al. Sympathetic outflow to muscle in humans during short periods of microgravity produced by parabolic flight. The American Journal of Physiology. 1999;**46**:R419-R426

[27] Miwa C, Mano T, Saito M, Iwase S, Matsukawa T, Sugiyams Y, et al. Ageing reduces symptaho-suppressive response to head-out water immersion in humans. Acta Physiologica Scandinavica. 1996;**158**:15-20

[28] Iwase S, Sugiyama Y, Miwa C, Kamiya A, Mano T, Ohira Y, et al. Effects of three days of dry immersion on muscle sympathetic nerve activity and arterial blood pressure in human. Journal of the Autonomic Nervous System. 2000;**79**:156-163

[29] Brunstetter T. Spaceflight Associated Neuro-ocular Syndrome (SANS). Current Clinical Insight & Questions of Interest. 2019. Available from: file:///C:/Users/owner/Desktop/Environmenal/brunstetter-sans-trish-red-risk-school-05apr18b-final.pdf

[30] Stenger MB. Spaceflight Associated Neuro-ocular Syndrome (SANS). 2019. Available from: file:///C:/Users/owner/Desktop/Environmenal/SANS%20by%20Stenger.pdf

[31] Koppelmans V, Bloomberg JJ, Mulavara AP, Seidler RD. Brain structural plasticity with spaceflight. npj Microgravity. 2016;**2**:2

[32] Van Ombergen A, Laureys S, Sunaert S, Tomilovskaya E, Parizel PM, Wuyts FL. Spaceflight-induced neuroplasticity in humans as measured by MRI: What do we know so far? npj Microgravity. 2017;**3**:2. DOI: 10, 1038/s1526-016-0010-8

[33] Iwase S. Effectiveness of centrifuge-induced artificial gravity with ergometric exercise as a countermeasure

during simulated microgravity exposure in humans. Acta Astronautica. 2005;**57**:75-80

[34] Nikawa T. Molecular mechanism and nutritional approach for unloading-mediated muscle atrophy. Journal of Japan Society of Nutrition and Food Sciences. 2017;**70**:3-8

[35] Powers SK, Kavazis AN, DeRuisseau KC. Mechanisms of disuse muscle atrophy role of oxidative stress. American Journal of Physiology. Regulatory, Integrative and Comparative Physiology. 2005;**288**:R337-R344

[36] Sandonà D, Desaphy JF, Camerino GM, Bianchini E, Ciciliot S, Danieli-Betto D, et al. Adaptation of mouse skeletal muscle to long-term microgravity in the MDS mission. PLoS One. 2012;**7**(3):e33232. DOI: 10.1371/journal.pone.0033232

[37] Uchida T, Abe T, Kohno S, Yamashita Y, Kaneko K, Kondo S, et al. Signal transduction of muscle cells under microgravity. Space Utilization Research. 2015;**29**:107-108

[38] DLR (German Aerospace Center), Institute of Aerospace Medicine, Research Report: Institute of Aerospace Medicine; 2017. pp. 1-45

[39] Planel H. Space and Life, an Introduction to Space Biology and Medicine. Boca Raton, FL: CRC Press; 1988. pp. 1-178

[40] Iwase S, Nishimura N, Mano T. Osteoporosis in spaceflight. In: Flores MV, editor. Topics in Osteoporosis. London, UK: IntechOpen; 2013. pp. 259-279

[41] Lang T, LeBlanc A, Evans H, Lu Y, Genant H, Yu A. Cortical and trabecular bone mineral loss from the spine and hip in long-duration spaceflight. Journal of Bone and Mineral Research. 2004;**19**:1006-1012

[42] Yamanaka JS, de Paiva MB, Carlos BL, Shimano AC. Effects of microgravity on bone. Aeronautics and Aerospace Open Access Journal. 2018;**2**:238-241

[43] LeBlanc AD et al. Skeletal responses to space flight and the bed rest analog: A review. Journal of Musculoskeletal & Neuronal Interactions. 2007;**7**:33-47

[44] Trudel G, Payne M, Mädler B, Ramachandran N, Lecompte M, Wade C, et al. Bone marrow fat accumulation after 60 days of bed rest persisted 1 year after activities were resumed along with hemopoietic stimulation: The women international space simulation for exploration study. Journal of Applied Physiology. 2009;**107**:540-548

[45] Spatz JM, Wein MN, Gooi JH, Qu Y, Garr JL, Liu S, et al. The Wnt inhibitor sclerostin is up-regulated by mechanical unloading in osteocytes in vitro. The Journal of Biological Chemistry. 2015;**290**:16744-16758. DOI: 10.1074/jbc.M114.628313

[46] Wolff J. The Law of Bone Remodeling. Berlin: Springer; 1986. pp. 1-120

[47] Kelly PJ, Bronk JT. Venous pressure and bone formation. Microvascular Research. 1990;**39**:364-375

[48] Kwon R, Meyas DR, Tang WJ, Frangos JA. Microfluidic enhancement of intramedullary pressure increases interstitial fluid flow and inhibits bone loss in hindlimb suspended mice. Journal of Bone and Mineral Research. 2010;**25**:1798-1807

[49] Prisby RD. Mechanical, hormonal and metabolic influences on blood vessels, blood flow and bone. The Journal of Endocrinology. 2017;**235**:R77-R100

[50] Qin Y, Kaplan T, Saldanha A, Rubin C. Fluid pressure gradients arising

from oscillations in intramedullary pressure, is correlated with the formation of bone and inhibition of intracortical porosity. Journal of Biomechanics. 2003;**36**:1427-1437

[51] Revell WJ, Brookes M. Haemodynamic changes in the rat femur and tibia following femoral vein ligation. Journal of Anatomy. 1994;**184**:625-633

[52] Naot D, Cornish J. The role of peptides and receptors of the calcitonin family in the regulation of bone metabolism. Bone. 2008;**43**:813-818. DOI: 10.1016/j.bone.2008.07.003

[53] Ikegame M, Rakopoulos M, Zhou H, Houssami S, Martin TJ, Moseley JM, et al. Calcitonin receptor isoforms in mouse and rat osteoclasts. Journal of Bone and Mineral Research. 1995;**10**:59-65

[54] Gooi JH, Pompolo S, Karsdal MA, Kulkarni NH, Kalajzic I, McAhren SH, et al. Calcitonin impairs the anabolic effect of PTH in young rats and stimulates expression of sclerostin by osteocytes. Bone. 2010;**46**:1986-1997

[55] O'Brien CA, Plotkin LI, Galli C, Goellner JJ, Gortazar AR, Allen MR, et al. Control of bone mass and remodeling by PTH receptor signaling in osteocytes. PLoS One. 2008;**3**(8):e2942. DOI: 10.1371/journal.pone.0002942

[56] Cox JF, Tahvanainen KU, Kuusela TA, Levine BD, Cooke WH, Mano T, et al. Influence of microgravity on astronauts' sympathetic and vagal responses to Valsalva's manoeuvre. The Journal of Physiology. 2002;**538**(Pt 1): 309-320

[57] Fu Q, Levine BD, Pawelczyk JA, Ertl AC, Diedrich A, Cox JF, et al. Cardiovascular and sympathetic neural responses to handgrip and cold pressor stimuli in humans before, during and after spaceflight. The Journal of Physiology. 2002;**544**(2):653-664

[58] Levine BD, Pawelczyk JA, Ertl AC, Cox JF, Zuckerman JH, Diedrich A, et al. Human muscle sympathetic neural and haemodynamic responses to tilt following spaceflight. The Journal of Physiology. 2002;**538**(Pt 1):331-340

[59] Iwase S, Mano T, Watanabe T, Saito M, Kobayashi F. Age-related changes of sympathetic outflow to muscles in humans. Journal of Gerontology. 1991;**46**:M1-M5

[60] Nishimura N, Iwase S, Shiozawa T, Sugenoya J, Shimizu Y, Takada M, et al. Effectiveness of countermeasure to bone metabolic deconditioning induced by simulated microgravity exposure (in Japanese). Space Utilization Research. 2010;**26**:122-124

[61] Robins SP, Woitge H, Hesley R, Ju J, Seyedin S, Seibel MJ. Direct, enzyme-linked immunoassay for urinary deoxypyridinoline as a specific marker for measuring bone resorption. Journal of Bone and Mineral Research. 1994;**10**:1643-1649

[62] Taaffe DR, Robinson TL, Snow CM, Marcus R. High-impact exercise promotes bone gain in well-trained female athletes. Journal of Bone and Mineral Research. 1997;**12**:255-260

[63] Grigoriev AI, Morukov BV, Oganov VS, Rakhmanov AS, Buravkova LB. Effect of exercise and bisphosphonate on mineral balance and bone density during 360 day antiorthostatic hypokinesia. Journal of Bone and Mineral Research. 1992;**7**(Suppl 2):S449-S455

[64] Rodan GA, Fleisch HA. Bisphosphonates: Mechanisms of action. The Journal of Clinical Investigation. 1996;**97**:2692-2696

[65] Thompson DD, Seedor JG, Weinreb M, Rosini S, Rodan GA. Aminohydroxybutane bisphosphonate inhibits bone loss due to immobilization

in rats. Journal of Bone and Mineral Research. 1990;**5**:279-286

[66] LeBlanc A, Shackelford L, Schneider V. Future human bone research in space. Bone. 1998;**22**(5 Suppl):113S-116S

[67] Watanabe Y, Ohshima H, Mizuno K, Sekiguchi C, Fukunaga M, Kohri K, et al. Intravenous pamidronate prevents femoral bone loss and renal stone formation during 90-day bed rest. Journal of Bone and Mineral Research. 2004;**19**:1771-1778

[68] LeBlanc A, Matsumoto T, Jones J, Shapiro J, Lang T, Smith SM, et al. Bisphosphonate as a countermeasure to space flight-induced bone loss. 2010. Available from: http://www.dsls.usra.edu/meetings/hrp2010/pdf/Bone/1094LeBlanc.pdf

[69] Durie BGM, Katz M, Crowley J. Osteonecrosis of the jaw and bisphosphonates. The New England Journal of Medicine. 2005;**353**:99-102

[70] Ruggiero SL, Mehrotra B, Rosenberg TJ, Engroff SL. Osteonecrosis of the jaws associated with the use of bisphosphonates: A review of 63 cases. Journal of Oral and Maxillofacial Surgery. 2004;**62**:527-534

[71] Ruggiero SL, Woo SB. Biophosphonate-related osteonecrosis of the jaws. Dental Clinics of North America. 2008;**52**:111-128

[72] De Santo NG, Cirillo M, Kirsch KA, Correale G, Drummer C, Frassl W, et al. Anemia and erythropoietin in space flights. Seminars in Nephrology. 2005;**25**:379-387

[73] Kunz H, Quiriarte H, Simpson BJ, Ploutz-Snyder R, McMonigal K, Sams C, et al. Alterations in hematologic indices during long-duration spaceflight. BMC Hematology. 2017;**17**:12-19

[74] Smith SM. Red blood cell and iron metabolism during space flight. Nutrition. 2002;**18**:864-866

[75] Alfrey CP, Udden MM, Leach-Huntoon C, Driscoll T, Pickett MH. Control of red blood cell mass in spaceflight. Journal of Applied Physiology. 1996;**81**:98-104

[76] Rizzo AM, Corsetto PA, Montorfano G, Milani S, Zava S, Tavella S, et al. Effects of long-term space flight on erythrocytes and oxidative stress of rodents. PLoS One. 2012;**7**(3):e32361

[77] Gunga H-C, Kirsch K, Baartz F, Maillet A, Gharib C, Nalishiti W, et al. Erythropoietin under real and simulated microgravity conditions in humans. Journal of Applied Physiology. 1996;**81**:761-773

[78] Crucian BE, Chouker A, Simpson R, Mehta S, Marshall G, Smith SM, et al. Immune system dysregulation during spaceflight: Potential countermeasures for deep space exploration missions. Frontiers in Immunology. 2018;**9**:1437. DOI: 10.3389/fimmu.2018.01437

[79] Clément G, Bukley A, editors. Artificial Gravity. Hawthorne, CA and New York, NY: Microcosm Press and Springer; 2007. pp. 1-364

[80] Young LR. Artificial gravity considerations for a Mars exploration mission. Annals of the New York Academy of Sciences. 1999;**871**:367-378

Chapter 3

Spaceflight-Associated Immune System Modifications

Jeremy Jeandel, Coralie Fonte, Gaetano Calcagno, Julie Bonnefoy, Stéphanie Ghislin, Sandra Kaminski and Jean-Pol Frippiat

Abstract

Spaceflight is an adverse environment characterized by a unique combination of stressors affecting almost all physiological systems, including the immune system. Indeed, several studies have shown that about 50% of the astronauts have faced immunological troubles. Here, we will review how spaceflight affects immune cell development, innate as well as adaptive immunity, required to ensure an efficient protection of the host, with a particular focus on T and B cells. Indeed, to better appreciate the risks associated to future long-duration space missions and to develop pharmacologic or nutritional countermeasures allowing immune system protection, it is mandatory to fully understand how these cell types are affected by space conditions. Finally, we will compare immune changes observed in astronauts with those encountered in the elderly, thereby illustrating the societal interest of space research.

Keywords: space exploration, immunity, lymphocytes, stress, aging

1. Introduction

Since Yuri Gagarin became the first human to leave the Earth's confines in 1961, more and more humans have traveled into space, and manned space stations have been built. During spaceflights, the organism is subjected to a variety of chronic and acute stressors. The first category comprises factors such as microgravity, confinement, isolation, radiation, and disturbed circadian rhythm. The second category covers periods of intense activity, such as spacewalks, but also hypergravity exposure during takeoff and landing. While acute stressors have been described as beneficial to the host as they can mobilize individual's defense capacities, several studies have shown that chronic stress has deleterious effects, as it contributes to the weakening of the immune system and the development of pathologies such as inflammatory disease, infections, and cancers [1–5]. In that context, it is interesting to note that 15 of the 29 astronauts involved in the Apollo missions developed bacterial or viral infections during, immediately after, or within a week of landing [6]. In addition, the very first epidemiological study based on medical data collected from 46 astronauts who spent 6 months on the ISS showed that 46% of them exhibited significant immunological problems [7]. Among notable events, 40% were classified as rashes/hypersensitivities and 27% as infectious diseases. Taken together, these data show that spaceflight-associated stressors affect, on average, the immune system of one out of two astronauts. Furthermore, these data demonstrate that immune system dysregulation occurs not only after landing but also during the flight [7].

In parallel of this immunological weakening, it is important to keep in mind that changes in microbial growth and pathogenicity have been observed [8, 9]. Depending on the bacteria studied, increased or decreased virulence [10, 11], altered sensitivity to antibiotics [12, 13], and/or increased biofilm formation [11, 13, 14] have been described as a result of the modulation of gene expression [9–11, 15, 16]. Moreover, there is some evidence to suggest that antibiotics may be less effective in space [12, 17, 18].

These microbial changes, combined with dysregulation of the immune system, certainly contribute to explain the increased susceptibility to infections observed in astronauts [19] (**Figure 1**). It is also noteworthy to keep in mind that, as the duration of space missions will increase, the potential for infectious diseases to arise during flight may become a critical issue because the probability of

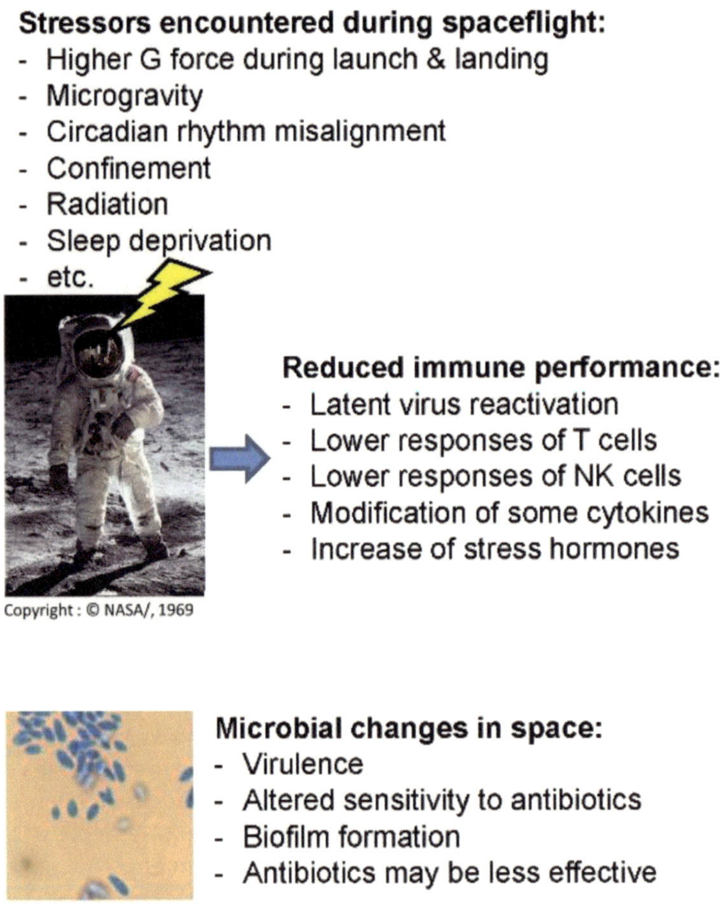

Figure 1.
Environmental changes associated with spaceflight (stressors) such as gravity change, the perturbation of the circadian rhythm (every day the residents of the ISS witness 16 sunrises and 16 sunsets), confinement, increased radiation, sleep deprivation, and nutritional factors weaken the immune system of about 50% of the astronauts. Most frequent immune changes consist in viral reactivations and lower responses of T and natural killer (NK) cells. These changes could be due to changes in cytokine expression and increased levels of stress hormones. In parallel, spaceflight environment might increase the virulence, resistance, and proliferation of some pathogens.

cross-contamination between crewmembers will increase. Additionally, microbial mutation rates may increase. Solar and cosmic radiation met during space missions, with a cumulative dose obviously increasing with mission duration, could contribute to the appearance of mutations potentially associated to resistance or diseases during such long and stressful endeavor as interplanetary missions.

Thus, caution should be paid to precisely understand how the immune system adapts, is modified/hampered, by unique environmental changes encountered during spaceflight. This knowledge is mandatory to allow the development of efficient biomedical strategies to preserve astronauts' heath during prolonged deep space exploration missions. In this chapter, we will review how spatial conditions affect the maturation of immune cells as well as the functions of mature immune cells required for the effective protection of the individual.

2. Effects on the maturation of immune cells

Cells that make up our immune system are derived from hematopoietic stem cells (HSC). These HSC will give rise to common myeloid progenitors (CMPs) and common lymphoid progenitors (CLPs). After several differentiation steps, CMPs will give raise to myeloid cells (granulocytes, monocytes, macrophages, and dendritic cells) and CLPs to lymphoid cells (B and T lymphocytes and NK cells). All of these cells are involved in natural and/or specific immunity.

A number of studies have analyzed the impact of spaceflight on the development of cells belonging to the myeloid lineage (or myelopoiesis). A decrease in the number of granulocyte and monocyte progenitors in rodents that have been in space or subjected to anti-orthostatic suspension (a model commonly used in the laboratory to reproduce many of the physiological changes observed in flight) has been demonstrated [20, 21]. The culture of human $CD34^+$ progenitors in flight has confirmed the inhibitory effect of microgravity on erythropoiesis (red blood cell production) [22]. Other studies have shown that the stressors encountered during spaceflight impact lymphocyte development (or lymphopoiesis). Diverse animal models have been used to address this question, such as mouse or the Iberian ribbed newt (*Pleurodeles waltl*, a urodele amphibian). The latter lends itself well to the constraints associated with space experiments and has all the cardinal elements of the mammalian immune system [23]. It has notably been observed that *P. waltl* larvae developed on board the ISS exhibit changes in the expression of IgM heavy-chain transcripts as well as a disruption in the expression of the Ikaros gene encoding transcription factors required for lymphopoiesis, suggesting that the latter could be weakened under spatial conditions [24]. This hypothesis was then confirmed in mice subjected to 21 days of anti-orthostatic suspension, which corresponds to a long-term mission at the human scale. It has been shown that this model induces a decrease in the number of CLPs and cells at the pro-B, pre-B, immature B, and mature B stages in the femoral bone marrow of suspended mice compared to control mice [25]. Furthermore, various causes of this weakening have been identified, such as a decrease in signal transduction by the interleukin-7 receptor and a decrease in the expression of transcription factors essential for B-cell development within the bone marrow. It has also been noted that this decrease in B lymphopoiesis is coupled with the remodeling of the bone tissue induced by the suspension, thereby reminding that all physiological systems interact within an organism and that these interactions have to be taken into account when analyzing the impact of stressors such as modeled microgravity. Finally, this sensitivity of hematopoiesis and the link with bone remodeling was confirmed in mice embarked on board the BION-M1 satellite for 30 days [26]. A decrease by a factor of two in the number of

B lymphocytes present in the bone marrow and a statistically significant decrease in the expression of factors required for the development of immune and bone cells were observed 7 days after returning to Earth but not on landing (**Figure 2**). This time delay can be explained by the fact that bone loss worsens after landing [27].

Note that in addition to the explanations presented above, glucocorticoids produced in response to chronic stress may contribute to altering B lymphopoiesis. Indeed, it has been demonstrated that continuous administration of corticosterone, via a subcutaneous implant, induces reprogramming of lymphopoiesis in mice, with a reduction of 30–70% of pro-B, pre-B, and immature B cells after 24 hours and a drop of 70–80% of pro-B and pre-B cells after 36 hours of treatment [29].

T lymphopoiesis (T-cell development in the thymus) is also affected by microgravity, as a decrease in T cells was observed in double-positive ($CD4^+CD8^+$) and single-positive ($CD4^+$ or $CD8^+$) maturation stages, when murine fetal thymuses were cultivated under simulated microgravity or spatial conditions [30]. This observation can be explained, at least in part, by the high sensitivity of thymocytes to stress [31]. Indeed, significant changes in mRNA expression from genes known to regulate stress and glucocorticoid receptors were observed in the thymus of mice subjected to a 13-day flight [32]. Another study did analyze the impact on murine T-cell antigen receptor (TCR) of being conceived and born under increased G force (2 G). This study revealed a disruption in TCR signaling and in the diversity of these receptor binding sites [33] (**Figure 3**) required for an individual to be able to specifically recognize peptides derived from the numerous antigens present in the

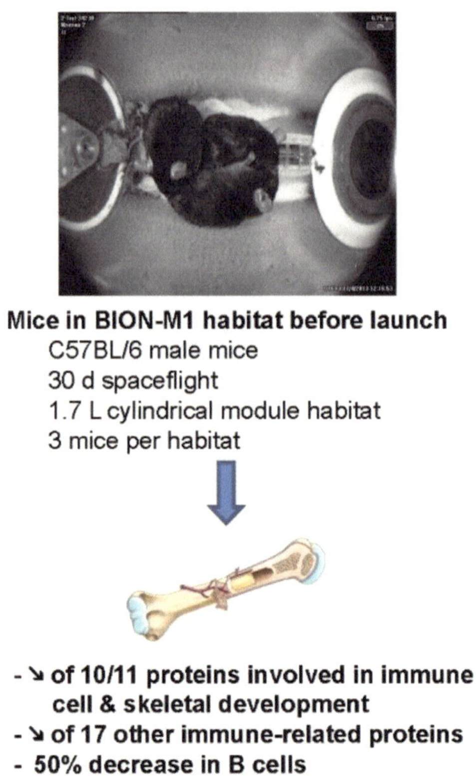

Figure 2.
Analysis of the femur from mice flown for 1 month on board the BION-M1 biosatellite revealed a decrease in the expression of 10 out of 11 proteins involved in immune cell and skeletal development, a decrease of the expression of 17 other immune-related proteins, and a 50% decrease in the number of B cells present in the bone marrow. Furthermore, this study showed that spaceflight effects were aggravated 1 week after landing [26]. Picture of mice in BION-M1 habitat from [28].

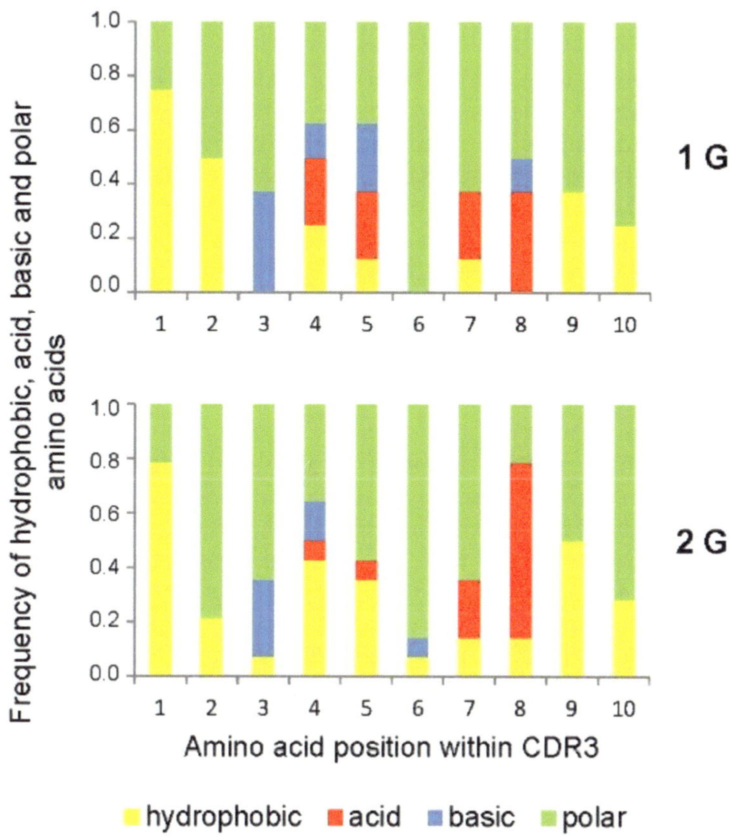

Figure 3.
Exposure to hypergravity during pregnancy affects TCR binding sites, thereby suggesting that the protection of the host might be affected [33]. T cells recognize an antigenic peptide on an MHC molecule at the surface of an antigen-presenting cell (APC) (dendritic cell, monocyte, macrophage, B cell). This recognition is ensured by the T-cell receptor (TCR) whose binding site is composed of six small polypeptide loops: two CDR1 loops, two CDR2 loops, and two CDR3 loops. CDR1 and CDR2 loops bind the MHC molecule. CDR3 loops bind the peptide. This figure presents the frequency of hydrophobic, acid, basic and polar amino acids at each position within TCR CDR3 loops from murine pups conceived and born at 1 or 2 G.

environment; 85% of the TCR repertoire was different in 2 G pups compared to control pups. Thus, the diversity of T-cell antigen receptor repertoire is significantly altered by 2 G exposure, which will likely affect host defense.

The impact of a model aiming at mimicking socio-environmental stresses experienced by astronauts [34] was then studied. This model involves the chronic exposure of mice to unpredictable socio-environmental stresses of various types (e.g., confinement, isolation, cage tilt, paired housing, perturbed circadian rhythm) and moderate intensity. It was demonstrated that this type of stressors only modifies 25% of the TCR repertoire [35]. Consequently, it appears that a change in the gravitational force has a much greater impact than socio-environmental stresses on the T-cell antigen receptor repertoire.

3. Effects on phagocytic and NK cells

Natural or innate immunity is the body's first line of defense against a pathogen after the skin and epithelial surfaces. It enables a non-specific response to be implemented, involving various types of immune cells such as neutrophils, monocytes, and macrophages, to destroy pathogens by phagocytosis and the release of microbicide substances.

Several studies have been conducted to understand how the space environment affects this immunity. For instance, an increase in the level of blood neutrophils in both humans and animals has often been observed after landing and can be attributed to the stresses encountered during this phase. Indeed, stress can induce the mobilization of these cells stored in the bone marrow [36, 37]. However, other explanations are also possible, such as changes in the expression of adhesion molecules [38]. It has also been shown that spatial conditions decrease the phagocytic and oxidative functions of neutrophils [39, 40] and induce, in monocytes, dysregulation in cytokine production, a reduced capacity to engulf *Escherichia coli* as well as lower reactive oxygen species (ROS) production and degranulation [41, 42]. Lower cytotoxicity of natural killer cells that provide immunological resistance and defense against foreign microorganisms but also against cells transformed because of, for example, a viral infection was observed [43, 44]. In addition, the reactivation of latent herpes viruses has frequently been reported. For example, Varicella-zoster virus (VZV) DNA has been detected in the saliva of astronauts during and immediately after a flight, while no VZV DNA was detected before launch [45]. Additional studies have revealed the presence of VZV in the saliva of 50% of astronauts during short spaceflights [46] and have shown that this percentage can increase up to 65% during long-duration missions [47]. Significantly, a few cases resulted in the development of shingles [45]. These viral reactivations are frequently coupled with

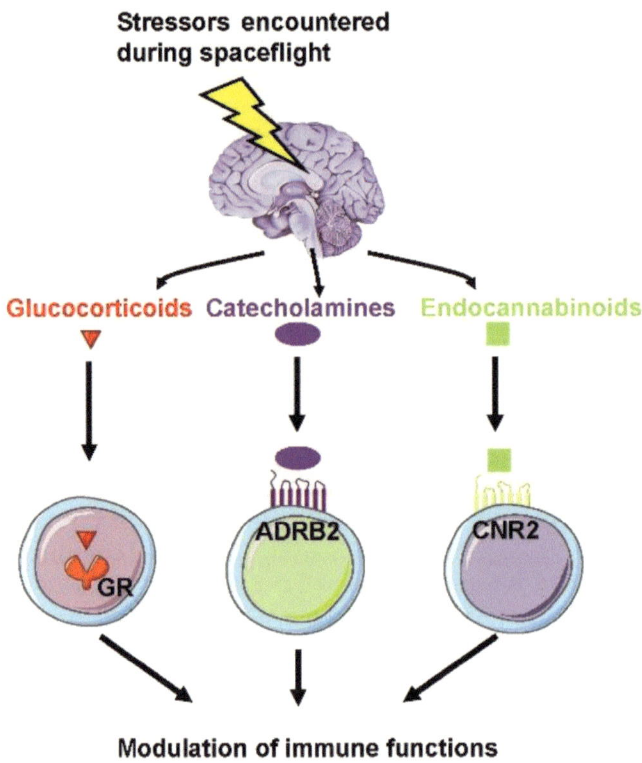

Figure 4.
Stressors encountered during space missions can induce the production of glucocorticoids, catecholamine, and endocannabinoids. Numerous immune cell types have receptors for these molecules. Their functions can therefore be directly affected by the binding of these molecules on these receptors. GR, glucocorticoid receptor; ADRB2, beta-2 adrenergic receptor; CNR2, cannabinoid receptor type 2.

a decrease in the production of interferons (cytokines constituting a first response in the event of viral infection) and to a higher level of stress hormones known to be able to regulate immune functions. Indeed, a variety of immune cells expresses glucocorticoid receptors, cannabinoid receptors, and adrenergic receptors (**Figure 4**). Thus, molecules produced in response to stressing events can directly affect immune cells and can be responsible for the reactivation of latent viruses [48–53].

Furthermore, virus reactivation could be a good biomarker of immunity weakening [54]. In support of this neuromodulation of the immune system, studies conducted on humans subjected to acute- (parabolic flight), medium- (1–2 weeks on board the ISS), or long-duration (4–7 months on board the ISS) gravitational stress demonstrated that there is a shift from an alert state of natural immune cells after acute gravitational stress to a decrease of their activity after spaceflight [55–57]. These changes were associated with changes in stress response, with a predominance of sympathetic nervous system responses after short flights, whereas long flights were characterized by glucocorticoid-induced changes. These data demonstrate that beside gravity change, stress responses are an important contributor to spaceflight-associated immune changes and once again highlight the importance of taking into account interconnections between physiological systems (here the nervous and immune systems).

4. Effects on antigen-presenting cells and lymphocytes

Specific or adaptive immunity is the second line of defense against the entry of foreign substances, particles, or cells into the organism. It involves natural and specific immune cells (antigen-presenting cells and lymphocytes) that will cooperate to develop a response specifically directed against the intruder.

APCs are a heterogeneous group that treat and present antigens in the form of peptides to $CD4^+$ T lymphocytes unable to recognize a native antigen via their TCR. These cells are crucial in triggering an immune response. This group includes dendritic cells, monocytes/macrophages, and B lymphocytes.

Even though the antigen presentation function is an essential immune process, very little information is available on the impact that environmental conditions encountered during spaceflights could have on this function. Only one study has been published on dendritic cells and revealed that microgravity reduces their production, their phagocytic capacities, and the surface expression of costimulatory/adhesion molecules involved in the presentation of antigenic peptides [58]. These data suggest that certain functions of antigen-presenting cells, required for the development of an effective immune response, may be disrupted in microgravity.

On the other hand, numerous studies have shown a significant reduction in T-cell activity under both real and simulated microgravity. This lower activity [59] results from spaceflight-induced modifications of the expression of genes essential for the proper functioning of T cells such as those encoding interleukin-2 and its receptor [60], translation of mRNAs [61], cell-cell interactions [62], alterations of the structure of the cytoskeleton [63–66], signal transduction enabling T-cell activation [67–69], and cell cycle regulation [70].

B lymphocytes are another cell type that acts in synergy with T lymphocytes to ensure optimal protection of the individual. These cells, at the maturation stage called plasmocyte, produce large quantities of antibodies, which, by binding specifically to the antigen, contribute to its elimination. Antibodies and B lymphocytes constitute humoral immunity whose modulation by spatial conditions has been much less studied than that of T lymphocytes. For many years, researchers have been satisfied with the quantification of antibodies present in the serum/

plasma of astronauts, but these studies generated conflicting results. For example, Konstantinova et al. [71] reported increased levels of serum IgA and IgM, while Rykova et al. [40] indicated that the amounts of serum IgA, IgG, and IgM were not affected after prolonged space missions. Subsequently, further studies were conducted to determine how changes in gravity affect humoral immunity and demonstrated that stresses encountered during spaceflight quantitatively and qualitatively affect the production of antibodies in response to antigenic stimulation. Changes in the expression of VH gene segments, encoding a large part of the antibody binding sites, have been observed in adult *P. waltl* immunized on board the Mir space station [72, 73] as well as a twofold decrease in the frequency of somatic hypermutations (SHM) that enable the diversification of antibody binding sites, in order to improve their affinity for the antigen [74] (**Figure 5**). Very recently, changes in the use of the gene segments required to create the antibody repertoire have also been observed in immunized mice subjected to anti-orthostatic suspension [75]. The antibody repertoire is therefore most likely modified under either real or simulated microgravity. In addition, a decrease in the expression of several effectors involved in immunity was observed 7 days after landing in mice that had been on board the BION-M1 biosatellite for 30 days [26]. This observation confirms the negative effect of spaceflight on the immune system and demonstrates that this impairment persists for at least 7 days after the return to Earth. This conclusion is in line with the studies that revealed disruptions in the production of antibodies in *P. waltl* still visible 10 days after landing [72–74].

Finally, it has been shown that the proliferative responses of B and T lymphocytes are reduced when mice are subjected to gravity changes (anti-orthostatic

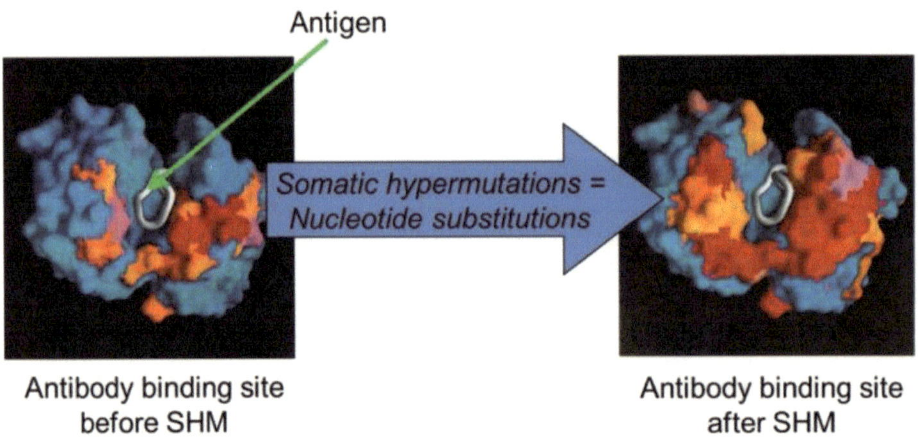

	SHM frequency (x 10^{-4}/pb)	P values ($\chi 2$ test)
Immunized on Earth (IE)	9.25	0.002 for IE/NI
Immunized in Space (IS)	4.65	0.031 for IS/NI

Figure 5.
Somatic hypermutations (SHM) are nucleotide substitutions whose purpose is to improve the affinity of antibody binding sites. The frequency of these mutations was determined in adult P. waltl immunized on board the Mir space station and in adult P. waltl immunized with the same antigen on Earth. This study showed that the frequency of these mutations is two times lower when animals are immunized on board the space station [74]. IE, P. waltl immunized on Earth; IS, P. waltl immunized in space; NI, not immunized.

suspension or 2 G hypergravity) for 3 weeks [76, 77]. However, the responses from these lymphocytes were not altered after 3 weeks of exposure to the model mimicking socio-environmental stressors encountered in flight [34]. These data suggest that the lower reactivity in lymphocytes induced by spaceflight is mainly due to gravity change.

Note that there is a break in the adaptation of mice at 3 G, which results in an increase in the serum corticosterone concentration and the level of anxiety [76]. These changes persist beyond 2 weeks after the return to normal gravity. This demonstrates that the hypergravity model should be used with caution if the effects of hypergravity are to be distinguished from those of a stress response. From 3 G, these two variables are cumulative.

5. Spaceflight as a model of accelerated immunosenescence

Certain immunological changes observed in astronauts or rodents on space missions can also be found in the elderly. For example, thymus involution, increased susceptibility to infections, and decreased response to vaccines may be correlated with impaired development of B- and T-lymphocyte function in the elderly [78, 79]. This thymus involution and changes in the development and response of immune cells are also observed when the gravitational force is altered, as illustrated by the reactivation of latent viruses in astronauts and the elderly. In addition, a recent study suggests that long-term spaceflight could induce an increase in inflammation as in the elderly (inflammaging), which could increase the risk of allergies or autoimmune diseases in astronauts [80]. Finally, aging is accompanied by changes in antibody production similar to those observed in flight. There is a decrease in antibody affinity [78] and a change in the use of antibody VH gene segments [81] as observed in *P. waltl* immunized in flight [73, 74], which affects the diversity of the antibody repertoire. It therefore appears that stresses encountered during space missions could lead to premature aging of the immune system.

6. Conclusion and perspectives

Studies conducted so far show that on average one out of two astronauts encounters immunological problems and that stressors encountered during spaceflights can affect all components of the immune system. It is therefore mandatory to understand in details how all immune cell types are affected by space conditions by unraveling the cellular and molecular mechanisms modified within these cells. Indeed, the impact of spatial conditions on certain cells and functions of the immune system have not yet been precisely determined. Furthermore, the impact of long-term missions is largely under-investigated. This is because, up to now, most scientific data are derived from space missions not exceeding 6 months in duration. In addition, the impact of spatial conditions on interconnections between the immune and other systems (such as the musculoskeletal, nervous, respiratory, and cardiovascular systems) should be studied using interdisciplinary approaches. All this knowledge is required (i) to gain a better understanding of the risks incurred during future long-duration space missions (such as planned mission to Mars), where the crew will be left to their own with no possibility of a rapid return to Earth, and (ii) to develop nutritional, psychosocial, and/or pharmacological countermeasures to reduce stress, preserve the immune system, and prevent the development or aggravation of diseases [82]. Another aspect that should be taken into account is in-flight monitoring of astronaut's health and diagnostic data using miniature and autonomous biosensors to help establish personalized treatments. This corresponds to a new field of research, space biotechnology, which aims to use advanced techniques ("omics" techniques) for the quantitative detection of proteins, nucleic acids, and metabolites in situ [83–86]. Such biosensors capable of analyzing minimum quantities of body fluids and of generating semiquantitative or quantitative results in a few minutes and with minimal resource consumption in terms of weight, volume, reagent storage, and energy will be required to allow deep space exploration. These researches and technological developments could also improve health on Earth as they could led to new therapeutic strategies to treat age- and stress-related immunosuppression and could likely contribute to improve point-of-care diagnostics at a patient's bedside, in a doctor's office, or hospital.

Acknowledgements

The authors acknowledge the support by the European Space Agency, the National Centre for Space Studies (CNES), the French Ministry of Higher Education and Research, the Université de Lorraine, the Région Lorraine, the Communauté Urbaine du Grand Nancy (CUGN), and the European Regional Development Fund (FEDER).

Conflict of interest

The authors declare no conflict of interest.

Author details

Jeremy Jeandel, Coralie Fonte, Gaetano Calcagno, Julie Bonnefoy, Stéphanie Ghislin, Sandra Kaminski and Jean-Pol Frippiat*
Faculty of Medicine, Stress Immunity Pathogens Unit, Lorraine University, Vandœuvre-lès-Nancy, France

*Address all correspondence to: jean-pol.frippiat@univ-lorraine.fr

References

[1] Dhabhar FS, McEwen BS. Acute stress enhances while chronic stress suppresses cell-mediated immunity in vivo: A potential role for leukocyte trafficking. Brain, Behavior, and Immunity. 1997;**11**:286-306. DOI: 10.1006/brbi.1997.0508

[2] Silberman DM, Wald MR, Genaro AM. Acute and chronic stress exert opposing effects on antibody responses associated with changes in stress hormone regulation of T-lymphocyte reactivity. Journal of Neuroimmunology. 2003;**144**:53-60. DOI: 10.1016/j.jneuroim.2003.08.031

[3] Godbout JP, Glaser R. Stress-induced immune dysregulation: Implications for wound healing, infectious disease and cancer. Journal of NeuroImmune Pharmacology. 2006;**1**:421-427. DOI: 10.1007/s11481-006-9036-0

[4] Webster Marketon JI, Glaser R. Stress hormones and immune function. Cellular Immunology. 2008;**252**:16-26. DOI: 10.1016/j.cellimm.2007.09.006

[5] Dhabhar FS. Enhancing versus suppressive effects of stress on immune function: Implications for immunoprotection and immunopathology. Neuroimmunomodulation. 2009;**16**:300-317. DOI: 10.1159/000216188

[6] Kimzey S. Hematology and immunology studies. In: Johnson RS, Dietlein LF, editors. Biomedical Results from Skylab. Washington DC: NASA; 1977. pp. 248-282

[7] Crucian B, Babiak-Vazquez A, Johnston S, Pierson DL, Ott CM, Sams C. Incidence of clinical symptoms during long-duration orbital spaceflight. International Journal of General Medicine. 2016;**9**:383-391. DOI: 10.2147/IJGM.S114188

[8] Horneck G, Klaus DM, Mancinelli RL. Space microbiology. Microbiology and Molecular Biology Reviews. 2010;**74**:121-156. DOI: 10.1128/MMBR.00016-09

[9] Zea L, Prasad N, Levy SE, Stodieck L, Jones A, Shrestha S, et al. A molecular genetic basis explaining altered bacterial behavior in space. PLoS One. 2016;**11**:e0164359. DOI: 10.1371/journal.pone.0164359

[10] Rosenzweig JA, Ahmed S, Eunson J, Chopra AK. Low-shear force associated with modeled microgravity and spaceflight does not similarly impact the virulence of notable bacterial pathogens. Applied Microbiology and Biotechnology. 2014;**98**:8797-8807. DOI: 10.1007/s00253-014-6025-8

[11] Cervantes JL, Hong BY. Dysbiosis and immune dysregulation in outer space. International Reviews in Immunology. 2016;**35**(1):67-82. DOI: 10.3109/08830185.2015.1027821

[12] Klaus DM, Howard HN. Antibiotic efficacy and microbial virulence during space flight. Trends in Biotechnology. 2006;**24**:131-136. DOI: 10.1016/j.tibtech.2006.01.008

[13] Lynch SV, Mukundakrishnan K, Benoit MR, Ayyaswamy PS, Matin A. *Escherichia coli* biofilms formed under low-shear modeled microgravity in a ground-based system. Applied and Environmental Microbiology. 2006;**72**:7701-7710. DOI: 10.1128/AEM.01294-06

[14] Kim H, Bhunia AK. Secreted listeria adhesion protein (lap) influences lap-mediated listeria monocytogenes paracellular translocation through epithelial barrier. Gut Pathogens. 2013;**5**:16. DOI: 10.1186/1757-4749-5-16

[15] Crabbé A, Schurr MJ, Monsieurs P, Morici L, Schurr J, Wilson JW, et al.

Transcriptional and proteomic responses of Pseudomonas aeruginosa PAO1 to spaceflight conditions involve Hfq regulation and reveal a role for oxygen. Applied and Environmental Microbiology. 2011;**77**:1221-1230. DOI: 10.1128/AEM.01582-10

[16] Shi J, Wang Y, He J, Li P, Jin R, Wang K, et al. Intestinal microbiota contributes to colonic epithelial changes in simulated microgravity mouse model. The FASEB Journal. 2017;**31**:3695-3709. DOI: 10.1096/fj.201700034R

[17] Juergensmeyer MA, Juergensmeyer EA, Guikema JA. Long-term exposure to spaceflight conditions affects bacterial response to antibiotics. Microgravity Science and Technology. 1999;**12**:41-47. PMID: 11543359

[18] Taylor PW, Sommer AP. Towards rational treatment of bacterial infections during extended space travel. International Journal of Antimicrobial Agents. 2005;**26**:183-187. DOI: 10.1016/j.ijantimicag.2005.06.002

[19] Frippiat JP, Crucian BE, de Quervain DJF, Grimm D, Montano N, Praun S, et al. Towards human exploration of space: The THESEUS review series on immunology research priorities. NPJ Microgravity. 2016;**2**:16040. DOI: 10.1038/npjmgrav.2016.40

[20] Dunn CD, Johnson PC, Lange RD, Perez L, Nessel R. Regulation of hematopoiesis in rats exposed to antiorthostatic, hypokinetic/hypodynamia: I. model description. Aviation. Space and Environmental Medicine. 1985;**56**:419-426. PMID: 4004676

[21] Ichiki AT, Gibson LA, Jago TL, Strickland KM, Johnson DL, Lange RD, et al. Effects of spaceflight on rat peripheral blood leukocytes and bone marrow progenitor cells. Journal of Leukocyte Biology. 1996;**60**:37-43. DOI: 10.1002/jlb.60.1.37

[22] Davis TA, Wiesmann W, Kidwell W, Cannon T, Kerns L, Serke C, et al. Effect of spaceflight on human stem cell hematopoiesis: Suppression of erythropoiesis and myelopoiesis. Journal of Leukocyte Biology. 1996;**60**:69-76. DOI: 10.1002/jlb.60.1.69

[23] Frippiat JP. Contribution of the urodele amphibian pleurodeles waltl to the analysis of spaceflight-associated immune system deregulation. Molecular Immunology. 2013;**56**:434-441. DOI: 10.1016/j.molimm.2013.06.011

[24] Huin-Schohn C, Guéguinou N, Schenten V, Bascove M, Gauquelin-Koch G, Baatout S, et al. Gravity changes during animal development affect IgM heavy-chain transcription and probably lymphopoiesis. The FASEB Journal. 2013;**27**:333-341. DOI: 10.1096/fj.12-217547

[25] Lescale C, Schenten V, Djeghloul D, Bennabi M, Gaignier F, Vandamme K, et al. Hind limb unloading, a model of spaceflight conditions, leads to decreased B lymphopoiesis similar to aging. The FASEB Journal. 2015;**29**:455-463. DOI: 10.1096/fj.14-259770

[26] Tascher G, Gerbaix M, Maes P, Chazarin B, Ghislin S, Antropova E, et al. Analysis of femurs from mice embarked on board BION-M1 biosatellite reveals a decrease in immune cell development, including B cells, after 1 wk of recovery on earth. The FASEB Journal. 2019;**33**:3772-3783. DOI: 10.1096/fj.201801463R

[27] Gerbaix M, White H, Courbon G, Shenkman B, Gauquelin-Koch G, Vico L. Eight days of earth reambulation worsen bone loss induced by 1-month spaceflight in the major weight-bearing ankle bones of mature mice. Frontiers

in Physiology. 2018;**9**:746. DOI: 10.3389/fphys.2018.00746

[28] Andreev-Andrievskiy A, Popova A, Boyle R, Alberts J, Shenkman B, Vinogradova O, et al. Mice in Bion-M 1 space mission: Training and selection. PLoS One. 2014;**9**:e104830. DOI: 10.1371/journal.pone.0104830

[29] Laakko T, Fraker P. Rapid changes in the lymphopoietic and granulopoietic compartments of the marrow caused by stress levels of corticosterone. Immunology. 2002;**105**:111-119. DOI: 10.1046/j.1365-2567.2002.01346.x

[30] Woods CC, Banks KE, Gruener R, DeLuca D. Loss of T cell precursors after spaceflight and exposure to vector-averaged gravity. The FASEB Journal. 2003;**17**:1526-1528. DOI: 10.1096/fj.02-0749fje

[31] Taves MD, Hamden JE, Soma KK. Local glucocorticoid production in lymphoid organs of mice and birds: Functions in lymphocyte development. Hormones and Behavior. 2017;**88**:4-14. DOI: 10.1016/j.yhbeh.2016.10.022

[32] Lebsack TW, Fa V, Woods CC, Gruener R, Manziello AM, Pecaut MJ, et al. Microarray analysis of spaceflown murine thymus tissue reveals changes in gene expression regulating stress and glucocorticoid receptors. Journal of Cellular Biochemistry. 2010;**110**:372-381. DOI: 10.1002/jcb.22547

[33] Ghislin S, Ouzren-Zarhloul N, Kaminski S, Frippiat JP. Hypergravity exposure during gestation modifies the TCRβ repertoire of newborn mice. Scientific Reports. 2015;**5**:9318. DOI: 10.1038/srep09318

[34] Gaignier F, Legrand-Frossi C, Stragier E, Mathiot J, Merlin JL, Cohen-Salmon C, et al. A model of chronic exposure to unpredictable mild socio-environmental stressors replicates some spaceflight-induced immunological changes. Frontiers in Physiology. 2018;**9**:514. DOI: 10.3389/fphys.2018.00514

[35] Fonte C, Kaminski S, Vanet A, Lanfumey L, Cohen-Salmon C, Ghislin S, et al. Socioenvironmental stressors encountered during spaceflight partially affect the murine TCR-β repertoire and increase its self-reactivity. The FASEB Journal. 2019;**33**:896-908. DOI: 10.1096/fj.201800969R

[36] Michurina TV, Domaratskaya EI, Nikonova TM, Khrushchov NG. Blood and clonogenic hemopoietic cells of newts after the space flight. Advances in Space Research. 1996;**17**:295-298. DOI: 10.1016/0273-1177(95)00650-4

[37] Guéguinou N, Huin-Schohn C, Bascove M, Bueb JL, Tschirhart E, Legrand-Frossi C, et al. Could spaceflight-associated immune system weakening preclude the expansion of human presence beyond Earth's orbit? Journal of Leukocyte Biology. 2009;**86**:1027-1038. DOI: 10.1189/jlb.0309167

[38] Stowe RP, Sams CF, Mehta SK, Kaur I, Jones ML, Feeback DL, et al. Leukocyte subsets and neutrophil function after short-term spaceflight. Journal of Leukocyte Biology. 1999;**65**:179-186. DOI: 10.1002/jlb.65.2.179

[39] Kaur I, Simons ER, Castro VA, Ott CM, Pierson DL. Changes in neutrophil functions in astronauts. Brain, Behaviour, and Immunity. 2004;**18**:443-450. DOI: 10.1016/j.bbi.2003.10.005

[40] Rykova MP, Antropova EN, Larina IM, Morukov BV. Humoral and cellular immunity in cosmonauts after the ISS missions. Acta Astronautica. 2008;**63**:697-705. DOI: 10.1016/j.actaastro.2008.03.016

[41] Crucian B, Stowe R, Quiriarte H, Pierson D, Sams C. Monocyte phenotype and cytokine production profiles are dysregulated by short-duration spaceflight. Aviation, Space, and Environmental Medicine. 2011;**82**:857-862. DOI: 10.3357/ASEM.3047.2011

[42] Kaur I, Simons ER, Kapadia AS, Ott CM, Pierson DL. Effect of spaceflight on ability of monocytes to respond to endotoxins of gram-negative bacteria. Clinical and Vaccine Immunology. 2008;**15**:1523-1528. DOI: 10.1128/CVI.00065-08

[43] Meshkov D, Rykova M. The natural cytotoxicity in cosmonauts on board space stations. Acta Astronautica. 1995;**36**:719-726. DOI: 10.1016/0094-5765(95)00162-X

[44] Taylor GR, Janney RP. In vivo testing confirms a blunting of the human cell-mediated immune mechanism during space flight. Journal of Leukocyte Biology. 1992;**51**:129-132. DOI: 10.1002/jlb.51.2.129

[45] Mehta SK, Cohrs RJ, Forghani B, Zerbe G, Gilden DH, Pierson DL. Stress-induced subclinical reactivation of varicella zoster virus in astronauts. Journal of Medical Virology. 2004;**72**:174-179. DOI: 10.1002/jmv.10555

[46] Mehta SK, Laudenslager ML, Stowe RP, Crucian BE, Sams CF, Pierson DL. Multiple latent viruses reactivate in astronauts during space shuttle missions. Brain, Behaviour, and Immunity. 2014;**41**:210-217. DOI: 10.1016/j.bbi.2014.05.014

[47] Mehta SK, Laudenslager ML, Stowe RP, Crucian BE, Feiveson AH, Sams CF, et al. Latent virus reactivation in astronauts on the international space station. NPJ Microgravity. 2017;**3**:11. DOI: 10.1038/s41526-017-0015-y

[48] Meehan R, Whitson P, Sams C. The role of psychoneuroendocrine factors on spaceflight-induced immunological alterations. Journal of Leukocyte Biology. 1993;**54**:236-244. DOI: 10.1002/jlb.54.3.236

[49] Crucian BE, Cubbage ML, Sams CF. Altered cytokine production by specific human peripheral blood cell subsets immediately following space flight. Journal of Interferon and Cytokine Research. 2000;**20**:547-556. DOI: 10.1089/10799900050044741

[50] Mehta SK, Stowe RP, Feiveson AH, Tyring SK, Pierson DL. Reactivation and shedding of cytomegalovirus in astronauts during spaceflight. The Journal of Infectious Diseases. 2000;**182**:1761-1764. DOI: 10.1086/317624

[51] Stowe RP, Mehta SK, Ferrando AA, Feeback DL, Pierson DL. Immune responses and latent herpesvirus reactivation in spaceflight. Aviation, Space, and Environmental Medicine. 2001;**72**:884-891 11601551

[52] Stowe RP, Pierson DL, Barrett AD. Elevated stress hormone levels relate to Epstein-Barr virus reactivation in astronauts. Psychosomatic Medicine. 2001;**63**:891-895. DOI: 10.1097/00006842-200111000-00007

[53] Mehta SK, Crucian BE, Stowe RP, Simpson RJ, Ott CM, Sams CF, et al. Reactivation of latent viruses is associated with increased plasma cytokines in astronauts. Cytokine. 2013;**61**:205-209. DOI: 10.1016/j.cyto.2012.09.019

[54] Cohrs RJ, Mehta SK, Schmid DS, Gilden DH, Pierson DL. Asymptomatic reactivation and shed of infectious varicella zoster virus in astronauts. Journal of Medical Virology. 2008;**80**:1116-1122. DOI: 10.1002/jmv.21173

[55] Stowe RP, Sams CF, Pierson DL. Effects of mission duration on

neuroimmune responses in astronauts. Aviation, Space, and Environmental Medicine. 2003;**74**:1281-1284. PMID: 14692473

[56] Crucian BE, Stowe RP, Pierson DL, Sams CF. Immune system dysregulation following short- vs long-duration spaceflight. Aviation, Space, and Environmental Medicine. 2008;**79**:835-843. DOI: 10.3357/ASEM.2276.2008

[57] Kaufmann I, Schachtner T, Feuerecker M, Schelling G, Thiel M, Choukèr A. Parabolic flight primes cytotoxic capabilities of polymorphonuclear leucocytes in humans. European Journal of Clinical Investigation. 2009;**39**:723-728. DOI: 10.1111/j.1365-2362.2009.02136.x

[58] Savary CA, Grazziuti ML, Przepiorka D, Tomasovic SP, McIntyre BW, Woodside DG, et al. Characteristics of human dendritic cells generated in a microgravity analog culture system. In Vitro Cellular and Developmental Biology—Animal. 2001;**37**:216-222. DOI: 10.1007/BF02577532

[59] Cogoli A, Tschopp A, Fuchs Bislin P. Cell sensitivity to gravity. Science. 1984;**225**:228-230. DOI: 10.1126/science.6729481

[60] Walther I, Pippia P, Meloni MA, Turrini F, Mannu F, Cogoli A. Simulated microgravity inhibits the genetic expression of interleukin-2 and its receptor in mitogen-activated T lymphocytes. FEBS Letters. 1998;**436**:115-118. DOI: 10.1016/s0014-5793(98)01107-7

[61] Hughes-Fulford M, Chang TT, Martinez EM, Li CF. Spaceflight alters expression of microRNA during T-cell activation. The FASEB Journal. 2015;**29**:4893-4900. DOI: 10.1096/fj.15-277392

[62] Cogoli-Greuter M, Meloni MA, Sciola L, Spano A, Pippia P, Monaco G, et al. Movements and interactions of leukocytes in microgravity. Journal of Biotechnology. 1996;**47**:279-287. DOI: 10.1016/0168-1656(96)01380-6

[63] Lewis ML, Reynolds JL, Cubano LA, Hatton JP, Lawless BD, Piepmeier EH. Spaceflight alters microtubules and increases apoptosis in human lymphocytes (Jurkat). The FASEB Journal. 1998;**12**:1007-1018. DOI: 10.1096/fasebj.12.11.1007

[64] Cogoli-Greuter M. Effect of gravity changes on the cytoskeleton in human lymphocytes. Gravitational and Space Biology Bulletin. 2004;**17**:27-37

[65] Meloni MA, Galleri G, Pippia P, Cogoli-Greuter M. Cytoskeleton changes and impaired motility of monocytes at modelled low gravity. Protoplasma. 2006;**229**:243-249. DOI: 10.1007/s00709-006-0210-2

[66] Meloni MA, Galleri G, Pani G, Saba A, Pippia P, Cogoli-Greuter M. Space flight affects motility and cytoskeletal structures in human monocyte cell line J-111. Cytoskeleton. 2011;**68**:125-137. DOI: 10.1002/cm.20499

[67] Boonyaratanakornkit JB, Cogoli A, Li CF, Schopper T, Pippia P, Galleri G, et al. Key gravity-sensitive signaling pathways drive T cell activation. The FASEB Journal. 2005;**19**:2020-2022. DOI: 10.1096/fj.05-3778fje

[68] Chang TT, Walther I, Li CF, Boonyaratanakornkit J, Galleri G, Meloni MA, et al. The Rel/NF-κB pathway and transcription of immediate early genes in T cell activation are inhibited by microgravity. Journal of Leukocyte Biology. 2012;**92**:1133-1145. DOI: 10.1189/jlb.0312157

[69] Martinez EM, Yoshida MC, Candelario TLT, Hughes-Fulford M. Spaceflight and simulated microgravity cause a significant reduction of key gene

expression in early T-cell activation. American Journal of Physiology. Regulatory, Integrative and Comparative Physiology. 2015;**308**:R480-R488. DOI: 10.1152/ajpregu.00449.2014

[70] Thiel CS, Paulsen K, Bradacs G, Lust K, Tauber S, Dumrese C, et al. Rapid alterations of cell cycle control proteins in human T lymphocytes in microgravity. Cell Communication and Signaling: CCS. 2012;**10**:1. DOI: 10.1186/1478-811X-10-1

[71] Konstantinova IV, Rykova MP, Lesnyak AT, Antropova EA. Immune changes during long-duration missions. Journal of Leukocyte Biology. 1993;**54**:189-201. DOI: 10.1002/jlb.54.3.189

[72] Boxio R, Dournon C, Frippiat JP. Effects of a long-term spaceflight on immunoglobulin heavy chains of the urodele amphibian pleurodeles waltl. Journal of Applied Physiology. 2005;**98**:905-910. DOI: 10.1152/japplphysiol.00957.2004

[73] Bascove M, Huin-Schohn C, Guéguinou N, Tschirhart E, Frippiat JP. Spaceflight-associated changes in immunoglobulin VH gene expression in the amphibian pleurodeles waltl. The FASEB Journal. 2009;**23**:1607-1615. DOI: 10.1096/fj.08-121327

[74] Bascove M, Guéguinou N, Schaerlinger B, Gauquelin-Koch G, Frippiat JP. Decrease in antibody somatic hypermutation frequency under extreme, extended spaceflight conditions. The FASEB Journal. 2011;**25**:2947-2955. DOI: 10.1096/fj.11-185215

[75] Rettig TA, Bye BA, Nishiyama NC, Hlavacek S, Ward C, Pecaut MJ, et al. Effects of skeletal unloading on the antibody repertoire of tetanus toxoid and/or CpG treated C57BL/6J mice. PLoS One. 2019;**14**:e0210284. DOI: 10.1371/journal.pone.0210284

[76] Guéguinou N, Bojados M, Jamon M, Derradji H, Baatout S, Tschirhart E, et al. Stress response and humoral immune system alterations related to chronic hypergravity in mice. Psychoneuroendocrinology. 2012;**37**:137-147. DOI: 10.1016/j.psyneuen.2011.05.015

[77] Gaignier F, Schenten V, De Carvalho Bittencourt M, Gauquelin-Koch G, Frippiat JP, Legrand-Frossi C. Three weeks of murine hindlimb unloading induces shifts from B to T and from th to tc splenic lymphocytes in absence of stress and differentially reduces cell-specific mitogenic responses. PLoS One. 2014;**9**:e92664. DOI: 10.1371/journal.pone.0092664

[78] Sasaki S, Sullivan M, Narvaez CF, Holmes TH, Furman D, Zheng NY, et al. Limited efficacy of inactivated influenza vaccine in elderly individuals is associated with decreased production of vaccine-specific antibodies. The Journal of Clinical Investigation. 2011;**121**:3109-3119. DOI: 10.1172/JCI57834

[79] Shaw AC, Goldstein DR, Montgomery RR. Age-dependent dysregulation of innate immunity. Nature Reviews. Immunology. 2013;**13**:875-887. DOI: 10.1038/nri3547

[80] Buchheim JI, Matzel S, Rykova M, Vassilieva G, Ponomarev S, Nichiporuk I, et al. Stress related shift toward Inflammaging in cosmonauts after long-duration space flight. Frontiers in Physiology. 2019;**10**:85. DOI: 10.3389/fphys.2019.00085

[81] Gibson KL, Wu YC, Barnett Y, Duggan O, Vaughan R, Kondeatis E, et al. B-cell diversity decreases in old age and is correlated with poor health status. Aging Cell. 2009;**8**:18-25. DOI: 10.1111/j.1474-9726.2008.00443.x

[82] Crucian BE, Choukèr A, Simpson RJ, Mehta S, Marshall G,

Smith SM, et al. Immune system dysregulation during spaceflight: Potential countermeasures for deep space exploration missions. Frontiers in Immunology. 2018;**9**:1437. DOI: 10.3389/fimmu.2018.01437

[83] Castro-Wallace SL, Chiu CY, John KK, Stahl SE, Rubins KH, McIntyre ABR, et al. Nanopore DNA sequencing and genome assembly on the international Space Station. Scientific Reports. 2017;**7**:18022. DOI: 10.1038/s41598-017-18364-0

[84] Karouia F, Peyvan K, Pohorille A. Toward biotechnology in space: High-throughput instruments for in situ biological research beyond earth. Biotechnology Advances. 2017;**35**:905-932. DOI: 10.1016/j.biotechadv.2017.04.003

[85] Roda A, Mirasoli M, Guardigli M, Zangheri M, Caliceti C, Calabria D, et al. Advanced biosensors for monitoring astronauts' health during long-duration space missions. Biosensors and Bioelectronics. 2018;**111**:18-26. DOI: 10.1016/j.bios.2018.03.062

[86] Zangheri M, Mirasoli M, Guardigli M, Di Nardo F, Anfossi L, Baggiani C, et al. Chemiluminescence-based biosensor for monitoring astronauts' health status during space missions: Results from the international Space Station. Biosensors and Bioelectronics. 2019;**129**:260-268. DOI: 10.1016/j.bios.2018.09.059

Chapter 4

Oral Tissue Responses to Travel in Space

Maija I. Mednieks and Arthur R. Hand

Abstract

The oral cavity functions in taste, mastication, solubilization and digestion of nutrients, as well as in respiration and speech, and participates in innate and adaptive immunity. Saliva creates and regulates the environment of the oral cavity, and changes in its composition and rate of secretion have significant effects on oral tissues as well as on systemic health. The effects of microgravity on the salivary glands, mandible and teeth were studied in mice flown on US space shuttle STS-131 and STS-135 missions, and the Russian Bion-M1 biosatellite. Significant changes in morphology and secretory protein expression occurred in parotid glands; submandibular glands were affected only on the 30-day Bion-M1 mission, indicating tissue specificity of the effects due to changes in gravity which may be similar to those taking place in humans. Changes also occurred in mandibular bone and incisor teeth. Collection of saliva is a non-invasive procedure for assessing physiological status and diagnosis of several disorders and provides a simple method for monitoring astronaut health during extended spaceflight.

Keywords: salivary glands, mandible, teeth, morphology, protein expression

1. Introduction

The oral cavity is the body's portal for the intake and initial processing of food and liquids, functioning in taste, mastication, solubilization and digestion of nutrients. Additionally, the oral cavity plays a role in respiration and speech as well as in innate and adaptive immunity. The components of the oral cavity include the teeth, consisting of three types of mineralized tissue, and their supporting structures, the bones of the maxilla and mandible and alveolar processes, and the periodontal ligament. The oral mucosa, which covers the alveolar processes, tongue, palate, cheeks, inside of the lips and floor of the mouth, consists of keratinized and non-keratinized epithelium overlying the connective tissues of the lamina propria and submucosa.

Located extraorally are the major salivary glands: the parotid, the submandibular and the sublingual, which convey their product, saliva, to the oral cavity *via* long ducts. Saliva, consisting of water, electrolytes and a large number of proteins, glycoproteins and small organic substances, creates and regulates the environment of the oral cavity. Changes in composition and rate of secretion of saliva have significant effects on oral tissues as well as on general health. In addition to the major glands, numerous small minor salivary glands are present in the subepithelial connective tissue throughout much of the oral cavity.

Among the organic constituents of saliva are digestive enzymes, calcium binding proteins, a variety of growth factors and regulatory molecules, antimicrobial components and immunoglobulins, and mucins that lubricate and moisten the oral tissues [1]. Other protective components of saliva include those involved in buffering and neutralizing acids produced by oral microorganisms and ingested with food and drink. Most are produced and secreted by the cells of the salivary glands, nevertheless a number of other proteins derived from other cells, tissues and organs also find their way into saliva. The presence of these substances and the ease of, and the noninvasive means of, collecting saliva have led to a great deal of interest for its use as a diagnostic fluid. Consequently, significant progress has been made in using saliva to detect oral and other cancers, several oral and systemic diseases, and to assess physiological and environmental stressors [2].

Our early studies of rats flown on Spacelab-3 [3, 4] indicated that exposure to microgravity resulted in specific changes in salivary gland structure and biochemistry. Accordingly, the premise for our recent studies of mice flown on the US space shuttles and the Russian Bion-M1 biosatellite, was that changes in the expression of salivary gland secretory proteins appear to occur in microgravity, and that the pattern of changes detectable in saliva could be employed to assess important aspects of the physiological status of astronauts.

In addition to studying salivary gland tissues, we examined the effects of microgravity on the mandibles and teeth of mice flown in space. Numerous studies have been carried out documenting the effects of spaceflight on bones of the weight-bearing skeleton, especially the bones of the lower limbs and vertebrae (see reviews in [5–7]). However, only a few studies have focused on non-weight-bearing bones, such as the mandible and cranial bones, and teeth [8, 9]. Teeth, once formed, are relatively stable structures, but can be affected by changes in the oral environment. An understanding of the effects of microgravity on tooth development, which can be studied using the continuously erupting rodent incisor, is important especially in view of future possibilities of long-term space journeys and colonization of other planets.

2. Effects of spaceflight on salivary glands

2.1 Morphology

The parenchyma of salivary glands consists of a collection of secretory end-pieces, or acini, connected to a ductal system that conveys saliva to the oral cavity [10, 11] (**Figure 1A**). The acini produce the primary saliva containing the majority of the proteins, most of the electrolytes and all of the water present in the final saliva. Serous acinar cells (**Figure 1B**) produce proteins and glycoproteins; seen in transmission electron micrographs (TEMs) the content of their secretory granules has a moderate to high density. Mucous acinar cells produce mainly mucins; their secretory granules exhibit a low density and often are fused with adjacent granules. The acinar secretory cells form a roughly spherical structure, their bases resting on the interstitial connective tissue, their apices facing a central lumen, and adjacent cells are joined by junctional complexes separating the interstitial from the luminal compartments. The cells contain abundant rough endoplasmic reticulum (RER), a prominent Golgi complex, and numerous secretory granules in the supranuclear cytoplasm. Secretory proteins synthesized by the cells are stored in the granules until their contents are released into the lumen by exocytosis. Consequently, analysis of the glandular tissue provides information about the composition of saliva.

Saliva produced in the acini flows into the initial portion of the ductal system, the intercalated ducts. These small ducts merge, forming larger ducts that

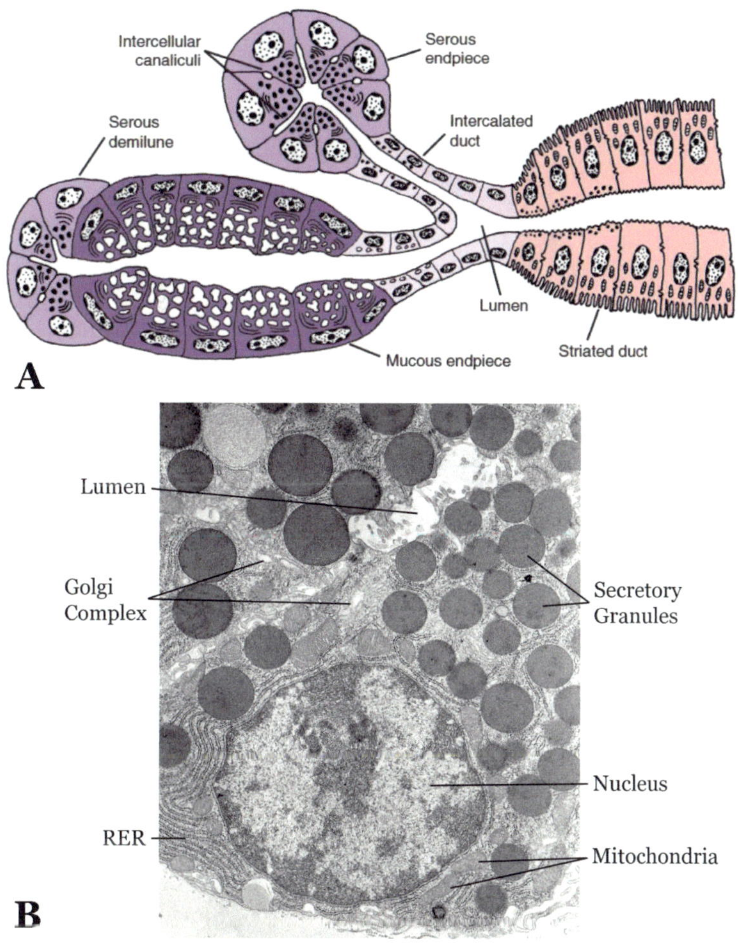

Figure 1.
(A) Diagram showing the main parenchymal components of mammalian salivary glands. (B) Transmission electron micrograph (TEM) of a serous acinar cell of a rodent parotid gland (PG) (modified, with permission, from ([10], Chapter 11, pp. 224, 225)).

eventually empty into the main ductal component, the striated ducts. The cells of these ducts are columnar in shape, and have abundant mitochondria nestled between infoldings of the basal cell membrane. Primary saliva produced by the acinar cells is modified by the striated (and excretory) ducts principally by reabsorption of Na^+ and Cl^- and addition of a few proteins; the final saliva is hypotonic with a pH near neutrality. The striated ducts become excretory ducts as they enter the connective tissue septa that partition the gland into lobules and lobes. The excretory ducts gradually merge into larger ducts and finally form the main excretory duct that empties into the oral cavity.

In humans, the acinar cells of the parotid gland (PG) are all serous. The submandibular gland (SMG) is a mixed gland, with mostly serous acini but also some mucous acini that are capped by serous cells forming a structure called a demilune. The sublingual gland (SLG) also is a mixed gland; its acini are predominantly mucous with serous demilunes. The acini and ductal systems of the rodent PG and SLG are very similar to those of the human (**Figure 2**). The structure of the rodent SMG (**Figure 3**) differs significantly from that of the human. The acini in rodents are all seromucous and produce a number of proteins and glycoproteins as well as mucins. As the animals become sexually mature, the cells of first portion of the striated ducts

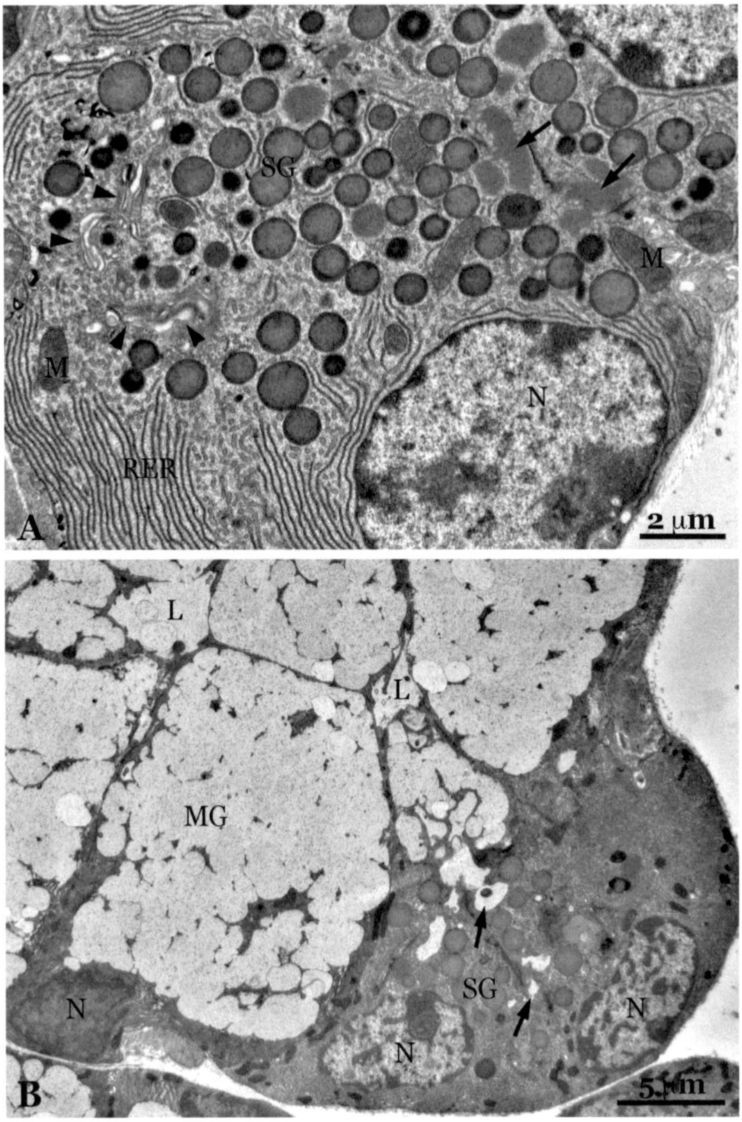

Figure 2.
TEMs of (A) PG serous acinar cells from a STS-135 flight mouse, and (B) sublingual gland (SLG) mucous acinar cells and serous demilune cells from a STS-131 habitat ground control mouse. Mucous acinar cells are filled with electron lucent mucous granules (MG). Serous demilune cells (lower right) contain electron dense secretory granules (SG). Lumen (L), mitochondria (M), nucleus (N), rough endoplasmic reticulum (RER), Golgi complex (arrowheads), intercellular canaliculi (arrows).

enlarge and synthesize a number of growth factors and proteases that are stored in large secretory granules in the apical cytoplasm. The development of this portion of the duct, called the granular convoluted duct, is strongly influenced by androgens; the granular convoluted duct is, therefore, more prominent in males than in females. Another feature of the sexual dimorphism seen in the rodent SMG is the presence in females of terminal tubule cells, or granular intercalated duct cells, at the acinar-intercalated duct junction. These cells are remnants of the early development of the gland; in males they are entirely eliminated by apoptosis by about 1 month of age.

Specific morphological changes were seen in the PG after a 13- to 15-day flight on the STS-135 and STS-131 shuttle missions, and 30 days on the Bion-M1 biosatellite [12, 13]. In the acinar cells autophagic vacuoles were common, and apoptotic

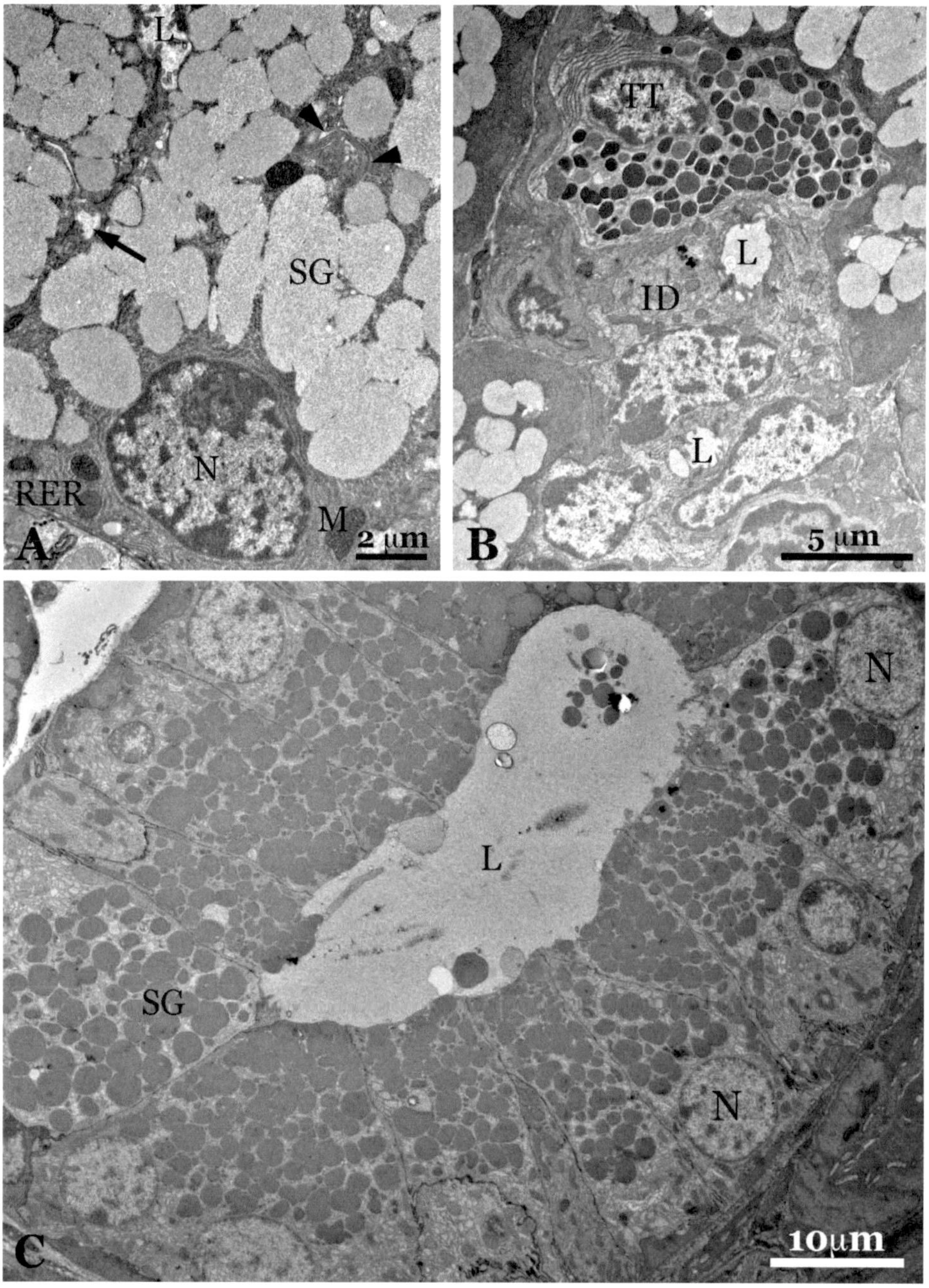

Figure 3.
TEMs of submandibular gland (SMG). (A) seromucous acinar cell from a STS-131 flight mouse, (B) intercalated duct and terminal tubule cell from a STS-131 habitat control mouse, and (C) granular convoluted duct from a Bion-M1 flight mouse. Acinar cells have a basal nucleus (N) and RER, and large secretory granules (SG) with a granular content of low density. In female mice, terminal tubule cells (TT) often are present in the intercalated ducts (ID) adjacent to the acinar cells. Columnar shaped granular convoluted duct cells of the male gland have a basal nucleus and mitochondria, and numerous dense secretory granules in the supranuclear cytoplasm. Lumen (L), mitochondria (M), Golgi complex (arrowheads), intercellular canaliculus (arrow).

cells were seen more frequently (**Figure 4A**). The autophagic vacuoles often contained degenerating secretory granules as well as other organelles. Some cells in the intercalated and striated ducts had large endocytic vacuoles containing dense content in the apical cytoplasm (**Figure 4B**). Immunogold labeling showed the presence of acinar secretory proteins in these vacuoles.

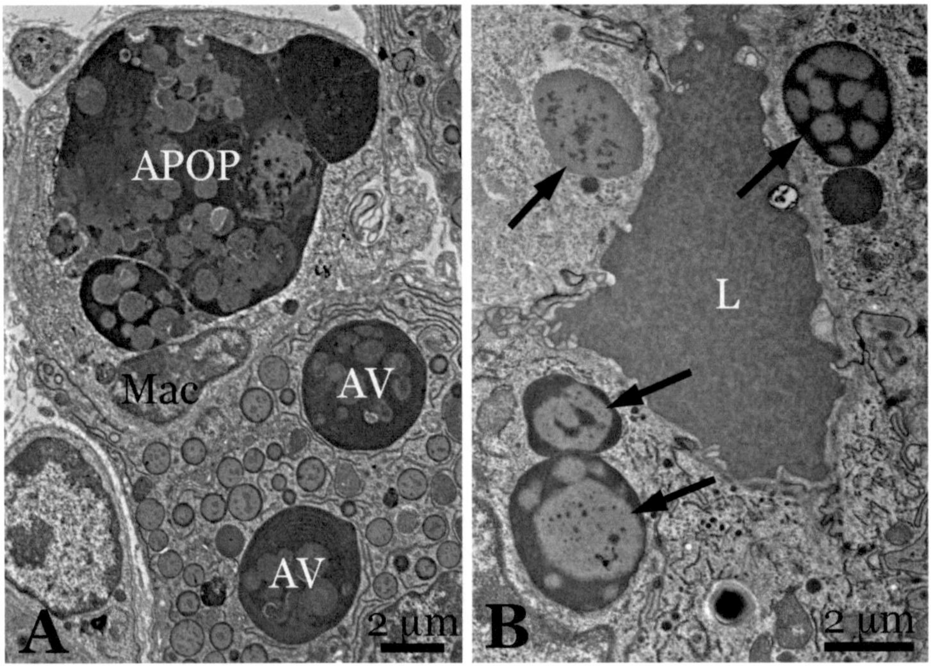

Figure 4.
TEMs of PG from STS-135 flight mice. (A) An acinar cell contains 2 autophagic vacuoles (AV), and a macrophage (Mac) has engulfed an apoptotic acinar cell (APOP). (B) Intercalated duct cells contain apical vacuoles (arrows) resulting from endocytosis of secretory proteins from the lumen (L) (with permission, from [13]).

The underlying basis for these changes is believed at least in part to be the loss of neural stimulation due to reduced masticatory activity. This is indicated by experiments where withholding food for 24–48 h, or feeding rats a liquid diet, results in numerous autophagic vacuoles containing degenerating secretory granules and an increase in the number of apoptotic cells in the PG [14–17].

The mice from the Bion-M1 mission all gained weight during the flight but were fed a soft paste diet. The mice from the space shuttle missions had *ad libitum* access to NASA rodent food bars [18], but lost a small amount of weight, as did the habitat control mice (maintained on the ground for the same length of time in the same habitats as the flight mice). This suggests that the mice on the space shuttle flights may not have eaten the amount of food sufficient to maintain their body weight. Since the number of autophagic vacuoles and frequency of apoptosis was greater in flight mice than in the habitat controls, microgravity also appears to affect these processes.

Endocytosis of salivary proteins by duct cells occurs in experimental diabetes [19, 20]. This suggests that the proteins may have an altered structure that is recognized as foreign by the duct cells; these cells are capable of endocytosing foreign proteins introduced in a retrograde fashion *via* the main excretory duct [21–23]. Alternatively, duct cell function in the flight mice may have been altered.

In contrast to the PG, no morphological changes were evident in the SMG or SLG of the flight mice. Similarly, feeding a liquid diet has no apparent effect on the morphology of the rat SMG and SLG [24].

2.2 Salivary protein expression studied using two major approaches

2.2.1 Electron microscopic immunogold labeling

Relative changes in secretory protein expression can be analyzed by electron microscopic immunogold labeling. Determination of the labeling density (gold particles/μm^2)

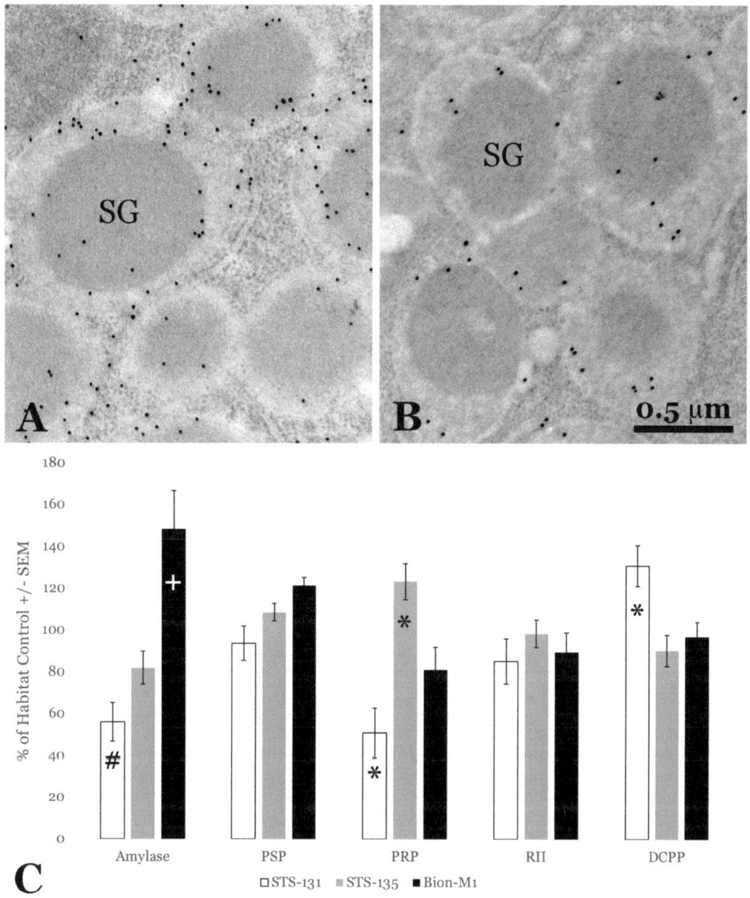

Figure 5.
*TEM immunogold labeling for proline-rich protein (PRP) in PG acinar cells of (A) flight and (B) habitat ground control mice from STS-135. Secretory granules (SG). (C) Quantitative TEM immunogold labeling of secretory proteins in PG of mice flown on STS-131 and STS-135, and the Bion-M1 biosatellite. α-Amylase, parotid secretory protein (PSP), PRP and the type II regulatory subunit of protein kinase A (RII) are present in acinar cell secretory granules; demilune cell and parotid protein (DCPP) is present in intercalated duct cell secretory granules. Labeling results are shown as a percentage of the corresponding habitat control mice ± SEM. +, $p = 0.065$; *, $p < 0.05$; #, $p < 0.01$.*

of secretory granules of PG acinar and duct cells showed alterations in the expression of several salivary proteins following spaceflight [12, 13] (**Figure 5**).

Compared to their habitat controls, expression of α-amylase, a digestive enzyme, was decreased in the parotid glands of mice flown on the space shuttles, but increased in mice flown on the Bion-M1 biosatellite. Parotid secretory protein (PSP), an antimicrobial protein [25, 26], was slightly increased in mice from the Bion-M1 flight. Proline-rich protein (PRP), a calcium and polyphenol binding protein [27, 28], was decreased in mice from STS-131, increased in mice from STS-135, and decreased, but not significantly, in mice from the Bion-M1 flight. The type II regulatory subunit of protein kinase A (PKA-RII), a stress marker secreted into saliva [29], was slightly decreased in mice from the 3 flights. Demilune cell and parotid protein (DCPP), secreted by intercalated duct cells and believed to have antimicrobial activity [30], was increased in mice from the STS-131 flight, but was unchanged in mice from STS-135 and Bion-M1 flights.

PKA-RII also is present in the nucleus, where it is involved in regulating several genes [31, 32]. No significant differences were seen in nuclear PKA-RII in parotid acinar cells of mice from the 3 flights (**Figure 6**).

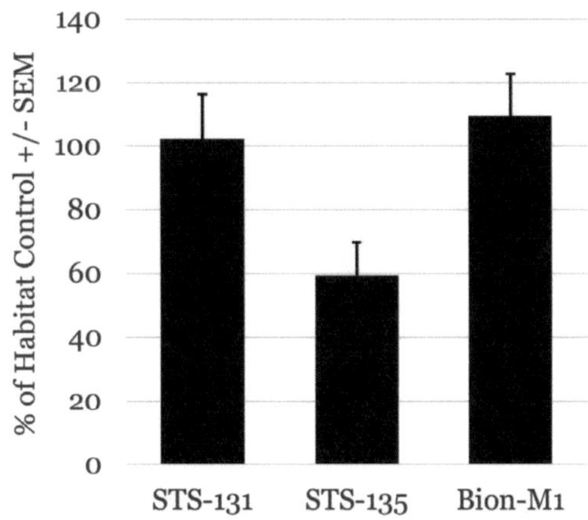

Figure 6.
Quantitative TEM immunogold labeling of nuclear PKA-RII in PG of mice flown on STS-131 and STS-135, and the Bion-M1 biosatellite. Labeling results are shown as a percentage of the corresponding habitat control mice ± SEM.

	STS-131	STS-135	Bion-M1
SABPα	114.6 ± 12.1	88.9 ± 26.3	221.9 ± 31.7*
PRP	111.1 ± 13.0	76.6 ± 9.0	134.4 ± 11.5
EGF	87.7 ± 12.4	36.3 ± 6.0	159.9 ± 4.6*
NGF	116.3 ± 23.3	89.9 ± 5.1	159.0 ± 10.6*
SMGC	83.0 ± 9.3	103.4 ± 5.3	—

Labeling results are shown as a percentage of the corresponding habitat control mice ± SEM.
$^{*}p < 0.01$.

Table 1.
Quantitative TEM immunogold labeling of secretory proteins in SMG of female mice flown on STS-131 and STS-135, and male mice flown on the Bion-M1 biosatellite.

The expression of SMG secretory proteins was essentially unchanged in mice from the 13–15-day space shuttle flights. However, mice from the Bion-M1 flight showed significantly increased expression of an acinar cell protein, salivary androgen binding protein alpha (SABPα), a pheromone involved in mate selection [33], and the granular convoluted duct cell proteins epidermal growth factor (EGF) and nerve growth factor (NGF) (**Table 1**). PRP, present in acinar cell secretory granules of both sexes, and submandibular gland protein C (SMGC), present in terminal tubule cell secretory granules of female mice, were not significantly different from controls. In the SLG, PSP expression by demilune cells was significantly increased in mice from the space shuttle STS-131 flight, but not in mice from STS-135 [34]. The expression of both PKA-RII in demilune cells, and the acinar cell mucin Muc19, were increased, but not significantly.

2.2.2 Electrophoresis and Western blotting

Polyacrylamide gel electrophoresis (PAGE) is the separation of proteins based on mobility on a gel matrix, subjected to a charge differential between two sealed chambers, of a body fluid or tissue extract. This technique is frequently used to determine the distribution or banding pattern of protein groups in tissues under different conditions. Quantification of the intensity of the bands and their relative mobility by PAGE is carried out by densitometric analysis. Saliva is a convenient

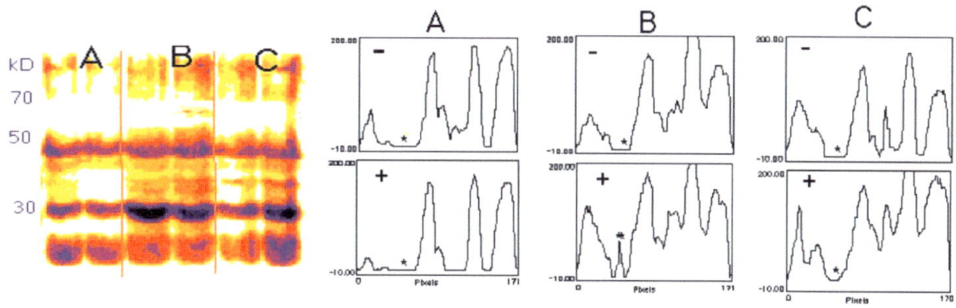

Figure 7.
Electrophoresis and densitometry of saliva from three randomly selected individuals (A, B, and C) from a pool of approximately 50 test subjects. The left ordinate on the gel pattern image shows molecular based migration rates of a mixture of proteins of known size. At the right of the polyacrylamide gel electrophoresis (PAGE) banding patterns are the corresponding densitometric measurements. The top row (−) prior to, the bottom row (+) after, chewing. The asterisks show the migration distance of RII that can be identified by Western blotting.

body fluid for measuring a variety of physiological conditions or responses on the basis of variations in protein content or distribution. For example, after testing more than 50 samples, PAGE separation of salivary proteins has shown that each individual pattern is unique. Namely, no superimposable patterns were found, therefore constituting essentially a salivary protein pattern fingerprint. Shown at the left in **Figure 7**, in duplicate, are the patterns of three (A, B, C) randomly selected individuals with the corresponding densitometric profiles at the right. **Figure 7** also shows how normal chewing changes the amplitude of several protein peaks, but not the basic pattern.

The PAGE protein banding patterns of the STS mission samples showed a decrease of some of the higher molecular weight bands while the faster moving smaller bands increased when these were compared to vivarium controls. The changes were not consistent between missions and may in part be due to lack of stability of the tissue or of sample preparation.

Western blotting is a technique using antibodies generated to a specific protein for its identification among numerous others in the banding pattern. Proteins are transferred onto nitrocellulose membranes and stained with a washable dye to show the band distribution, then quantified by densitometry. The dye is removed and the membrane probed for reactivity with a specific antibody. These techniques were employed to test salivary gland extracts to measure protein patterns of flight animal samples when compared with those of ground control animals (**Figure 8**).

The results show that the RII peaks in the STS-135 flight samples (**Figure 8**) were significantly smaller than the corresponding band amplitude in either vivarium or habitat controls, middle and bottom profiles, respectively. Western blotting experiments to determine the expression of α-amylase show a significant decrease in the flight and in the habitat controls.

These results indicate that on the shorter, STS-135 flight the expression of RII was decreased, while in the longer Bion-M1 flight (**Figure 9**) an apparent stabilization or adjustment to the microgravity environment had occurred and the RII levels were not different, perhaps even increased from those of either of the controls. Both Bion-M1 and STS-135 flight and both control samples show a significant, but opposite change of a faster moving band that may represent a small protein or peptide.

The findings regarding RII, generally thought to be a stress associated protein [35, 36] responding to the conditions of spaceflight [7], varied by methods of analysis but also between flights within a particular test measure from morphometric analyses of immunogold labeling experiments. Bion-M1 Western blotting results showed no difference between RII in Bion-M1 flight and vivarium control

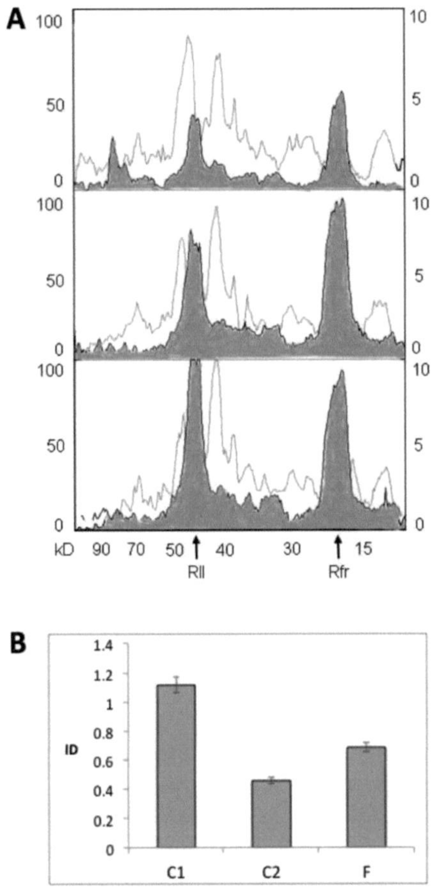

Figure 8.
Densitometric analysis of PAGE protein separation and anti-RII and anti-α-amylase reactivity in STS-135 mice. (A) PG protein samples from mice housed in, top panel, flight; middle panel, habitat; and bottom panel, vivarium type cages. The lighter curve represents the protein profile; the shaded area is the reactivity to anti-RII antibody. The ordinate axes are integrated density values of the proteins on the left-hand axis, and the integrated density values for RII and the reactivity of RII fragment (Rfr) on the right-hand axis. (B) PG α-amylase Western blotting. The ordinate shows integrated density, and the error bars show ±5% error. C1, vivarium control; C2, habitat control; F, flight. The values on the ordinate axis (ID) are integrated density in arbitrary units (with permission, from [13]).

mice. Measuring RII, therefore, may be a sensitive test for stress reactions during spaceflight.

The variability may be due to the sensitivity of responses affecting RII synthesis. Previous studies point to a rapid and transient increase in the cases of brief acute stress, for example lift off of a spacecraft, followed by long-term, perhaps low level, prolonged time under stress, decrease. For example, a 60-day to a year tour resulting in a decrease of RII and leveling off at concentrations below normal. Eventual physiologic adjustment to flight conditions may start to bring RII production toward normal. An apparent increase also may result from a rough landing that might have occurred in the Bion-M1 flight and the acute response obliterating any reduction due to chronic long-term flight conditions.

Tissue extract samples showed a large faster moving component, presumably an RII fragment. The fragments did resemble the densitometric changes in the flight versus the control. These differences did not completely account for the change of flight values compared to the control. There may have been smaller fragments not visible on the Western blot that could account for the decreases in flight RII.

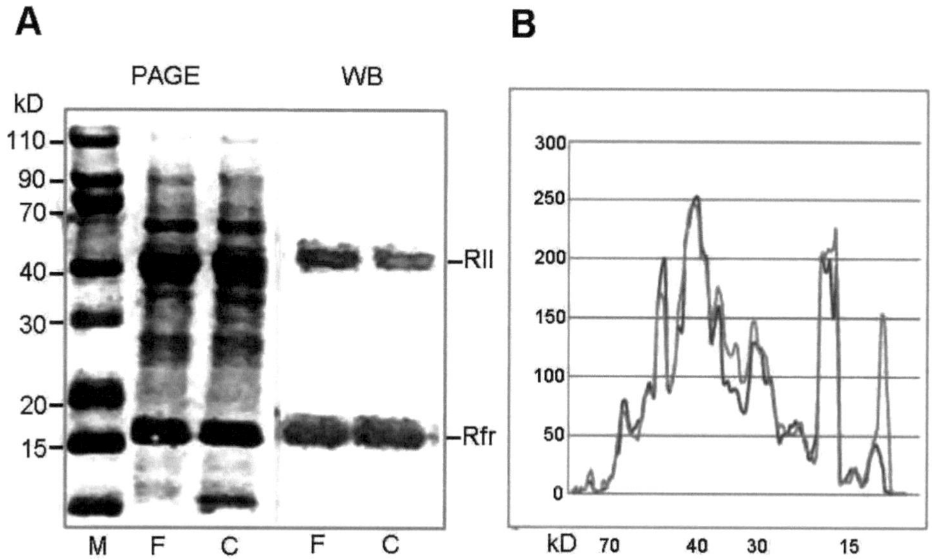

Figure 9.
Electrophoresis and Western blotting of flight (F) and control (C) PG extracts from mice of the Bion-M1 mission. (A) Protein banding patterns of the flight and control animals (F and C). The left lane (M) shows the relative mobilities of marker proteins of kilodalton (kD) size from 110 to 15 marked on the ordinate. Western blotting for RII is shown in the two right most lanes with two major bands, the first at an estimated 50 kDa size (RII), the second deemed to be a fragment between 14 and 20 kDa (Rfr). (B) densitometric profile of the control banding pattern which was not distinguishable from that of the flight (with permission, from [13]).

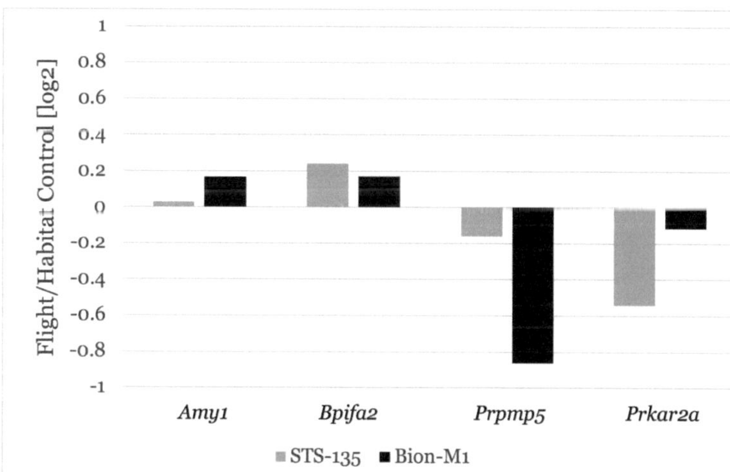

Figure 10.
Microarray analysis of secretory protein gene expression in PG of mice flown on STS-135 and the Bion-M1 biosatellite. Results are expressed as log_2 of the ratio of expression in flight mice to that in habitat control mice. Amy1, α-amylase; Bpifa2, PSP, Prpmp5, PRP; Prkar2a, RII.

2.3 Gene expression

Microarray analyses of the PG showed significant changes in the expression of numerous genes of flight mice compared to habitat ground control mice. In female mice flown on STS-135, the expression of 130 genes was significantly upregulated (log_2 ratio ≥ 1), and 75 genes were downregulated (log_2 ratio ≤ -1), (**Figure 10**). In male mice flown on Bion-M1, the expression of 70 genes was upregulated, and 65 genes were downregulated.

Interestingly, significant changes in gene expression occurred in the PG of habitat ground control mice compared to vivarium housed mice. In ground control female mice housed in the NASA Animal Enclosure Modules, the expression of 50 genes was significantly upregulated, and 33 genes were downregulated. For example, Western blotting of PG of STS-135 habitat control mice showed a decrease in the expression of α-amylase (**Figure 8B**). In ground control male mice housed in the Bion-M1 habitats, the expression of 42 genes was significantly upregulated, and 169 genes were downregulated. These results indicate that the effects of the habitats must be considered when assessing the overall effects of spaceflight on animal (as well as human) physiology.

Changes in the expression of secretory protein genes (**Figure 10**), for example, PKA-RII, largely paralleled the changes seen in protein expression as determined by immunogold labeling and Western blotting (e.g., PKA-RII expression in PG from STS-135 flight mice, **Figure 8B**).

There have been a few previous studies of salivary glands of rodents following spaceflight, and saliva collected from astronauts and cosmonauts before and after spaceflight. No changes in PG morphology were seen by light microscopy in mice flown for 12.5 days on Apollo 17 [37]. The SMG of rats flown for 18.5 days on Cosmos 936 and Cosmos 1129 showed a reduced glycoprotein content as determined by light microscopic histochemical staining [38, 39]. The saliva of 18 astronauts making up the primary and backup crews for Skylab missions of 28-, 59- and 84-days was analyzed preflight and postflight for flow rate, electrolytes, protein, immunoglobulin A (IgA) and lysozyme [40]. Changes were minimal except for an increase in IgA and a decrease in lysozyme concentrations. α-Amylase in cosmonaut saliva was decreased following an 18-day flight (uncited statement in [41]). The use of saliva for monitoring drug absorption was proposed, and saliva levels of acetaminophen were determined in 1 astronaut [42]. Analyses of saliva of cosmonauts during long-term spaceflight showed levels of cortisol below baseline [43], whereas salivary cortisol biorhythms and concentrations were preserved in astronauts onboard Spacelab [44] and cosmonauts on the International Space Station (ISS) [45]. Transient increases in saliva urea and phosphate concentrations, monitored in 2 astronauts during 6 months on the ISS, were consistent with serum concentrations reported in earlier studies [46]. Although studied in astronaut plasma, not saliva, in the NASA Twin Study, EGF and NGF levels were altered during a 340-day flight and after landing [47]. Our studies and those cited above indicate the feasibility of using saliva to monitor astronaut physiology and health.

3. Effects of spaceflight on the mandible

3.1 MicroCT analyses

MicroCT was used to assess mandibular bone volume (BV) and bone mineral density (BMD) in mice from the STS-135 and Bion-M1 flights [48].

In the STS-135 mission no differences in BV or BMD were seen between the flight and habitat ground control mice. However, when compared to vivarium control mice, BV was greater in both flight mice and habitat ground control mice. Similarly, BMD was greater in both the flight and habitat control mice than the vivarium control mice, although only the habitat control values were statistically significant. The increased BV and BMD seen in the flight and habitat control mice may be due to the difference in composition and consistency of the NASA food bars [18] these mice consumed compared to the standard rodent chow eaten by the vivarium control mice. The food bars have about 20% less caloric value per gram than chow, requiring greater consumption and more masticatory effort for comparable nutrition.

In contrast, mandibular BV decreased in both flight and habitat control mice compared to vivarium control mice from the Bion-M1 mission [48]. No differences occurred in BMD among these three groups, however.

These changes in flight and habitat control mice are likely due to the difference in diet between these two groups and the vivarium mice [49, 50]. The soft paste diet consumed by the Bion-M1 flight and habitat control mice required considerably less masticatory force than chow. Previous studies have shown that reduced mechanical loading results in decreased growth of the mandible [51, 52]. The different age and sex (9 week old females vs. 15–16 week old males, respectively) of the STS-135 and Bion-M1 mice also may have contributed to the different results for BV and BMD. The STS-135 mice were skeletally immature [53, 54], and female mice are physically more active than males [55] and have a different pattern of chewing and biting activity [56].

Earlier studies examined the effects of spaceflight and simulated weightlessness on mandibular bone. Periosteal osteogenesis was reduced in areas of the mandible not covered by muscle (molar region) in rats flown for 18.5 days on Cosmos 1129 [57]. Formation of alveolar bone was reduced on the modeling (mesial or anterior) side of the first molar, indicating a slowing of the normal distal (posterior) drift of rodent molars. There also was a decrease in alveolar bone mineral and collagen in fractions representing the most mature components, with a corresponding increase in the most immature fractions, suggesting a delay in maturation. The mandibles of rats flown for 12 days on Cosmos 1887 had relatively high Ca and Mg levels but otherwise a normal composition, although their hydroxyapatite crystals were smaller in size [58].

Studies of rodents subjected to hindlimb unloading (HU) for up to 4 weeks have been used as model of exposure to microgravity [5, 59, 60]. Although the effects on the spine and hindlimbs show similarities to changes seen in animals after spaceflight, effects on the mandible do not mimic those occurring after actual spaceflight. Ten to 14 days of HU did not alter maturation of mandibular bone matrix and mineral as seen in rats flown on Cosmos 1129 [61]. Similarly, 15 days of HU did not alter the total weight, ash weight, Ca content or Ca uptake of the mandible [62]. No significant effects on the mandible were seen after 28 days of HU [63]. An earlier study, however, found an increase in the dry and ash weights of the mandible after 3 weeks of HU [64]. In HU, the mandible still closes against gravity [60], thus spaceflight induced changes in the mandible are most likely a result of hypogravity [61].

Ghosh et al. [8] studied mandibles from mice flown on STS-131 and STS-135 (from the opposite side of the same STS-135 mice we studied [48]). They also found no difference in BV between the flight and habitat ground control mice from the 13-day STS-135 mission. However, mandibular BV was decreased in flight mice from the 15-day STS-131 mission. It is unlikely that the 2-day difference in flight length would result in a significant difference in BV. As noted by Ghosh et al., it is more likely that other factors are involved, such as the age difference and resulting skeletal maturation of the mice on the two flights (23 weeks vs. 9 weeks for STS-131 and STS-135, respectively), slight differences in food consumption, and/or the relatively low statistical power due to the small number of animals. They found no difference in BMD between the flight and habitat ground control mice from the STS-131 mission, but similar to our results, BMD was increased in the flight mice from the STS-135 mission. The factors mentioned above also could contribute to the differences in BMD seen between the mice on the two shuttle flights. The difference in BV between the Ghosh et al. study and our results for the STS-135 mice may be due to the different methods used to assess BV. Ghosh et al. measured BV in a single section (6 μm) from each mandible. In our study the results from 10 serial sections (120 μm span) per mandible were averaged, providing a more representative measurement.

3.2 Bone protein expression

The expression of the bone proteins sclerostin and osteocalcin in the mandibles of mice from the STS-135 and Bion-M1 missions was studied by immunohistochemistry (IHC) [48] (**Figure 11A–C**). Sclerostin is secreted by osteocytes and other cells embedded in mineralized tissues; it inhibits osteoblast proliferation and differentiation and induces osteoblast apoptosis [65, 66]. No differences in the intensity of sclerostin staining of osteocytes were observed between the flight mice and their respective controls in either mission. The percentage of reactive osteocytes, however, varied between the 2 missions and among the 3 groups and locations within the mandible. Overall, the percentage of reactive osteocytes was greater in the STS-135 mice than in the Bion-M1 mice, and greater in flight mice than in control mice. The alveolar bone anterior to the first molar and the interradicular bone of the first molar showed the greatest percentage of reactive osteocytes, whereas the septal bone between the first and second molars had the lowest percentage.

Sclerostin synthesis is increased by mechanical unloading (i.e., rodent HU, human bed rest) [67, 68]. However, the percentage of sclerostin-positive osteocytes was unaltered in male mouse tibiae after 3 or 7 days of HU [67]. Sclerostin-positive

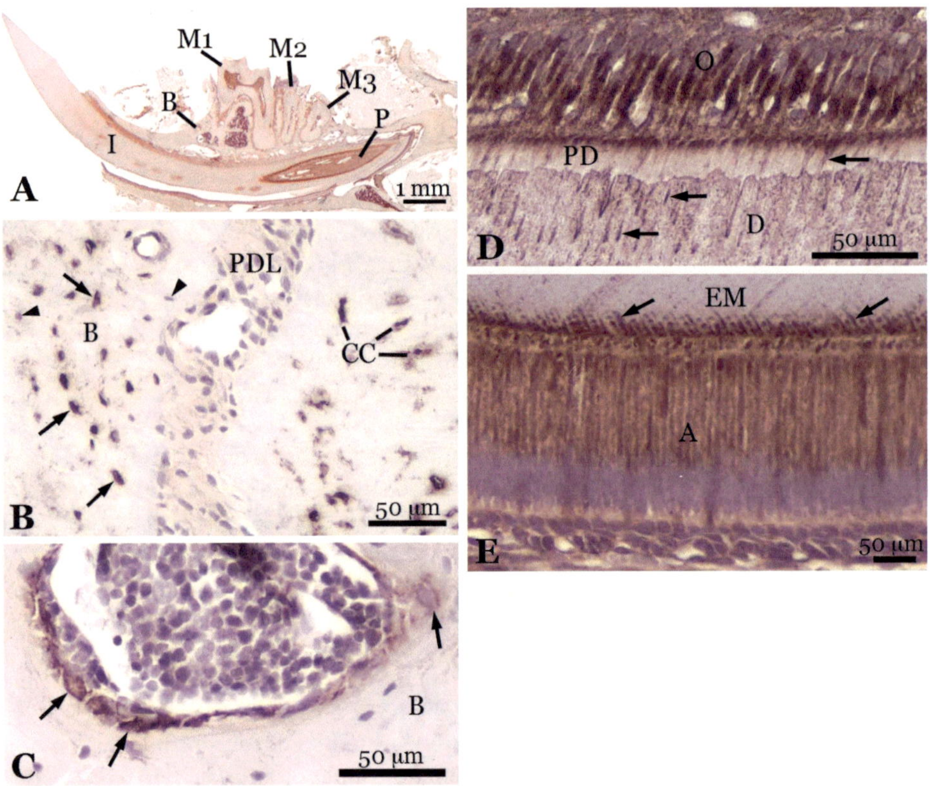

Figure 11.
Immunohistochemistry of mandibular bone and incisor teeth. (A) Sagittal section of a hemimandible from a Bion-M1 habitat control mouse, immunostained for osteocalcin. Bone (B), incisor (I), molars (M1-3), pulp (P). Note staining of odontoblasts lining the pulp chamber. (B) Sclerostin staining of osteocytes (arrows) in interseptal bone of a Bion-M1 flight mouse. A few osteocytes are unstained (arrowheads). Cementocytes (CC) in the cementum covering a molar root also stain for sclerostin. Periodontal ligament (PDL). (C) Osteocalcin staining of osteoblasts (arrows) of a Bion-M1 vivarium control mouse. (D) Immunostaining of dentin sialoprotein in odontoblasts (O) of a STS-135 flight mouse. Odontoblast processes (arrows) in predentin (PD) and dentin (D) are stained. Predentin and dentin stain less intensely. (E) Amelogenin staining of secretory ameloblasts (A) of a STS-135 vivarium control mouse. Tomes' processes of ameloblasts (arrows) are stained intensely. Enamel matrix (EM).

osteocytes increased in male rat femurs after 28 days of HU [69], and partial weight bearing (1/6G) for 21 days also increased their percentage in female mouse femurs [70]. No significant changes in serum sclerostin levels of astronauts and cosmonauts were observed after 4–6 months on the ISS [71, 72]. In the NASA Twins Study, a trend toward increased serum sclerostin levels was seen throughout the 340-day stay on the ISS [47].

Spaceflight also resulted in increases in the size of osteocyte lacunae, expression of matrix metalloproteinase (MMP) activity, and the percentage of osteocytes expressing tartrate-resistant acid phosphatase (TRAP) in pelvic bones of STS-131 flight mice, indicating osteocytic osteolysis [73]. Empty lacunae indicating osteocyte death doubled in number in the femoral cortex of Bion-M1 flight mice compared to vivarium and habitat control mice [74].

Osteocalcin is a non-collagenous bone matrix protein produced by osteoblasts (and odontoblasts). It binds calcium and has long been thought to regulate mineralization [75, 76]. More recently osteocalcin has been recognized as a major bone hormone that participates in the regulation of energy metabolism, brain development and cognition, and male fertility [77, 78]. No differences were observed in the IHC staining intensity of osteoblasts, bone matrix or osteoid between the flight mice and their respective controls from the STS-135 and Bion-M1 missions [48].

Osteocalcin synthesis in rat long bones and vertebrae decreased after short-term spaceflight (4–14 days) [79–82]. Serum and urinary osteocalcin levels increased in cosmonauts during 1 month and 6 months spaceflight [83–85]. A trend toward increased serum osteocalcin levels was seen in astronauts and cosmonauts after 4-6 months on the ISS [72], and during a 340-day flight on the ISS [47]. These findings are consistent with decreased bone formation and increased bone resorption. The lack of apparent change in osteocalcin staining in the mandible, along with the lack of or small changes in bone volume, suggests that the continued mastication of food prevents significant bone resorption.

Osteocalcin gene and protein expression also have been studied in cultured osteoblasts or osteoblast-like cells flown in space. Production of osteocalcin message after 9 days of microgravity by human osteosarcoma cells stimulated with vitamin D3 and transforming growth factor beta was only 19% of that of similarly treated cells at unit gravity [86]. The steady-state levels of osteocalcin mRNA in human fetal osteoblast cells after 17 days of microgravity did not differ from controls [87]. Osteocalcin protein expression was slightly but significantly increased in cultured human osteoblasts flown for 11 days [88]. The different results for osteocalcin expression may be due to the different cell types, different culture conditions, and different flight lengths.

The percentage of alveolar bone surface occupied by TRAP-positive osteoclasts (Oc.S./B.S. %) was increased in the mandibles of Bion-M1 flight and habitat control mice [48], indicating increased bone resorption. This correlates with the decreased BV observed in these mice compared to vivarium controls. Osteoclasts were present mainly along the mesial (anterior) surfaces of the alveolar bone, consistent with the normal physiological distal (posterior) drift of rodent molars. No differences were seen among the STS-135 mice groups, although the overall Oc.S./B.S. % was greater in these mice than in the Bion-M1 mice. This likely was due to the younger age of the STS-135 mice, as well as the increased proportion of sclerostin-positive osteocytes in these mice, as sclerostin promotes increased osteoclast formation and size [66]. In rats flown for 18.5 days on Cosmos 1129, there was a slight but non-significant decrease in alveolar bone resorption immediately after flight [57]. However, after 6 and 29 days of post-flight recovery, a significant decrease in bone formation and resorption was observed, indicating a slowing of the normal distal drift of the molars.

4. Effects of spaceflight on teeth

4.1 Incisor growth and development

The rodent mandible (and maxilla) contains, on each side, 1 incisor separated by a toothless region (diastema) from 3 molars (**Figure 11A**). The molars develop and erupt in a process similar to those of humans. The incisor, however, forms and erupts continuously, allowing the mandibular and maxillary incisors to maintain contact as the covering enamel and supporting dentin are worn away as the animal eats. The cells that form enamel, ameloblasts, originate from oral ectoderm; their stem cells are located in the cervical loop of the incisor, embedded within the bone of the mandible at the apical (posterior) end of the tooth. Ameloblasts initially deposit a partially mineralized matrix based mainly on the protein amelogenin. When the final thickness of the matrix is achieved, the ameloblasts undergo a morphologic and functional differentiation and begin to remove the matrix and add Ca and PO_4 until the enamel is fully mineralized. The cells that form dentin, odontoblasts, are derived from craniofacial ectomesenchyme. Throughout the life of the tooth, odontoblasts continually form and mineralize dentin, which has a collagen-based extracellular matrix. Odontoblasts also synthesize several non-collagenous proteins, including dentin sialoprotein (DSP), dentin phosphoprotein, osteocalcin, bone sialoprotein, osteopontin and dentin matrix protein-1 [89].

4.2 MicroCT analyses

Incisor teeth of flight and habitat ground control mice from the Bion-M1 mission had significantly greater enamel, dentin and overall volumes than the vivarium control mice (**Figure 12**). Enamel thickness in MicroCT sections at the 1st molar level was similar in all three mouse groups. In the flight and habitat control mice, however, the enamel had reached its full thickness at the 3rd molar level, and the thickness was greater in these 2 groups than in the vivarium control mice at both the 2nd and 3rd molar levels. Dentin thickness at all 3 molar levels in the flight and habitat control mice, as well as tissue density (mg hydroxyapatite/cm^3), was greater than that of the vivarium control mice. Additionally, as visualized in 3D reconstruction images, incisor mineralization in the flight and habitat control mice began further posteriorly than in vivarium control mice (**Figure 12G–H**). These observations are consistent with a decreased incisor eruption rate in the flight and habitat control mice, most likely due to the soft paste diet consumed by these two groups [90–92]. In the STS-135 mice, there was a trend toward an increased incisor volume in the flight mice compared to the vivarium control mice, but no differences in enamel and dentin volumes, thicknesses or tissue densities.

Previous studies of spaceflight effects on tooth development have produced differing/conflicting results. In some investigations there were no differences between flight and control animals in Ca and P content of incisor dentin [58, 93, 94]. In other studies Ca and P concentrations were increased [57, 95, 96], or decreased [97]. These varying results are likely due to differences in age and sex of the animals, diet, length of the flight, the region of the tooth examined, as well as the methods used in these investigations.

4.3 Tooth protein expression

The expression of osteocalcin, DSP, amelogenin and PKA-RII was evaluated by IHC in the tissues and cells of the teeth [48]. Odontoblasts and some cementocytes of molar roots expressed strong reactivity for osteocalcin (**Figure 11A**); weaker

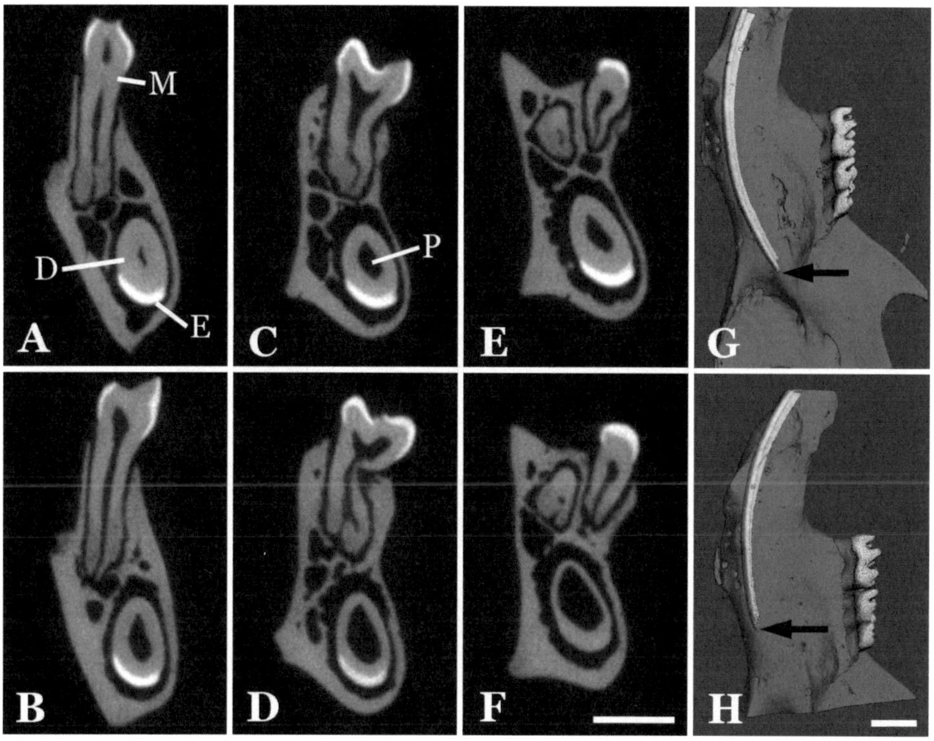

Figure 12.
MicroCT images of mandibles of flight (A, C, E, and G) and vivarium control (B, D, F, and H) mice from the Bion-M1 mission. Sections at the level of the 1st (A and B), 2nd (C and D) and 3rd (E and F) molars. Incisor dentin (D) and enamel (E), molar (M), pulp chamber (P). Note the earlier appearance of mineralized enamel (images E vs. F), thicker dentin and smaller central pulp chambers in the incisors of the flight mice. Three-dimensional reconstructions of the mandibles (G and H) show that incisor mineralization begins (arrows) further posterior in flight mice than in vivarium control mice. Habitat ground control mice showed the same features as flight mice. Scale bars = 1 mm.

reactivity was seen in predentin and dentin. Semiquantitative analysis revealed no differences in staining intensities among any of the flight and control groups. Immunostaining for DSP was strong in odontoblasts (**Figure 11D**) and some cementocytes; weaker reactivity was observed in predentin, dentin, molar cementum, enamel matrix, osteoblasts and bone. Significant differences in DSP staining intensities were observed among the Bion-M1 mouse groups. Predentin and cellular cementum (and bone) stained more intensely in flight mice than in vivarium controls. Secretory ameloblasts (**Figure 11E**) and enamel matrix (especially during enamel maturation) stained positively for amelogenin. Reactivity was significantly decreased in enamel matrix of Bion-M1 habitat control mice. Both odontoblasts and ameloblasts were immunoreactive for PKA-RII. Staining intensity was decreased in odontoblasts of Bion-M1 flight and habitat control mice compared to vivarium controls, and decreased in secretory ameloblasts of STS-135 flight and habitat control mice. In addition to effects due to microgravity, the observation of decreased amelogenin and PKA-RII reactivities in the habitat control mice suggest that the habitat environment may play a role in cellular and tissue responses.

5. Effects of spaceflight on other oral tissues

There are few reported studies of the effects of spaceflight on oral tissues other than bone and teeth. Histologic examination of periodontal tissues and oral mucosa revealed no differences between mice flown on Apollo 17 for 12.5 days and ground control mice [37]. Histologic studies of the lingual mucosa of rhesus monkeys after 30 days of simulated weightlessness showed no differences from control monkeys [98]. The masseter muscles of mice flown for 13 days on STS-135 showed no change in mass, fiber size distribution, signaling pathways, or the expression of genes associated with muscle atrophy compared to ground control mice [99]. Power output was decreased about 40% in the masseter muscles of flight mice, but maximal shortening velocity was not affected. In contrast, tibialis anterior muscles of these flight mice underwent atrophy, with loss of mass, changes in signaling pathways and gene expression, and loss of strength. In a separate experiment, mice fed a liquid diet for 2 weeks showed a decrease in masseter muscle fiber size of more than 40%. These results suggest that the load imposed by chewing hard food protected the masseter muscles from the microgravity-induced atrophy occurring in appendicular muscles, but was insufficient to preserve muscle power.

6. Conclusions

Oral tissues, particularly the salivary glands, mandible and teeth, are affected by spaceflight. The 3 major salivary glands respond differently to the effects of microgravity and the spaceflight habitats. The PG exhibits a number of morphological changes that are also seen following restriction of food, feeding of a liquid diet, and experimental diabetes. The expression of some secretory proteins is increased, whereas others are decreased; for some proteins the length of microgravity exposure may be important. A number of genes also exhibit increased or decreased expression. The response of the SMG to spaceflight appears to differ between males and females, with male glands exhibiting increases in expression of several secretory proteins and female glands remaining unaltered. The SLGs of female mice flown on the space shuttles showed a few changes in secretory protein expression. Interestingly, the expression of the stress protein PKA-RII and its gene was altered in the shorter space shuttle flights, but not the longer Bion-M1 flight. This may indicate an adaptation to the spaceflight environment, or a relatively rapid return to baseline during the longer period between landing and sample collection for the Bion-M1 mice. The ease of saliva collection makes it a potential alternative to blood for monitoring physiological responses to spaceflight and the health of astronauts and cosmonauts. Recent studies have documented substantial differences in male and female salivary gland gene expression in both mice [100] and humans [101]; therefore, additional, more comprehensive studies of the effects of spaceflight on the saliva proteome are needed.

Not considered in our studies, but important for overall oral health, are the minor salivary glands. Although they produce less than 10% of total saliva volume, these glands secrete continuously, providing mucins and antimicrobial substances that lubricate, moisten and protect the oral tissues, especially during periods when the major glands are inactive, such as during sleep. Microgravity induced alterations in their secretions could have significant effects on the teeth and oral mucosa.

The effects of spaceflight on the mandible, observed in our studies, are consistent with the observations reported by others for non-weight bearing skeletal elements. Changes in BV and bone architecture are minimal in the mandible compared to long bones of the lower extremity and vertebrae. The decrease in BV seen in the Bion-M1 flight and habitat control mice most likely was due to the soft paste diet consumed by these mice. Microgravity probably also has an effect, considering the increased proportion of sclerostin-positive osteocytes and number of osteoclasts in the mandible seen in the flight mice. Similarly, the significant changes observed in incisor tooth volume and mineralization appear to be related to the soft diet and slowing of the eruption rate. Changes in tooth protein expression may be a combined result of microgravity, diet and the habitat environment. The change from "normal" conditions, the confined space, and different diet, must also be considered, and adequately controlled for in experimental studies.

Many of the reported animal and human studies, including ours, would be improved with a larger number of subjects, increasing statistical power and strengthening observed trends. Nevertheless, many significant changes have been documented, and should be considered in planning future animal studies and especially long-term human spaceflight. The recent NASA Twins Study [47] emphasizes the need for comprehensive studies of long-term exposure to the spaceflight environment. Advances in spacecraft and experimental design will resolve current limitations and improve outcomes of future travel in space.

Acknowledgements

The authors' studies were supported by NASA (grant number NNX09AP13G [MIM]), the Connecticut Space Grant College Consortium (grant numbers P-536 and P-777 [ARH]), NIH (Shared Instrumentation Grant number 1S10RR024742 [ARH]) and UConn Health.

Conflict of interest

The authors have no conflicts of interest.

Author details

Maija I. Mednieks and Arthur R. Hand*
UConn Health, Farmington, Connecticut, USA

*Address all correspondence to: hand@uchc.edu

References

[1] Dawes C, Pederson AML, Villa A, Ekström J, Proctor GB, Vissink A, et al. The functions of human saliva: A review sponsored by the World Workshop on Oral Medicine VI. Archives of Oral Biology. 2015;**60**:863-874

[2] Dawes C, Wong DTW. Saliva and salivary diagnostics in the advancement of oral health. Journal of Dental Research. 2019;**98**:133-141

[3] Mednieks MI, Hand AR. Biochemical and morphological evaluation of the effects of space flight on rat salivary glands. The Physiologist. 1985; **28**(6 Suppl):S215-S216

[4] Mednieks MI, Hand AR. Salivary gland ultrastructure and cyclic AMP dependent reactions in Spacelab 3 rats. American Journal of Physiology. Regulatory, Integrative and Comparative Physiology. 1987;**252**:R233-R239

[5] Nagaraja MP, Risin D. The current state of bone loss research: Data from spaceflight and microgravity simulators. Journal of Cellular Biochemistry. 2013;**114**:1001-1008

[6] Orwoll ES, Adler RA, Amin S, Binkley N, Lewiecki EM, Petak SM, et al. Skeletal health in long-duration astronauts: Nature, assessment, and management recommendations from the NASA Bone Summit. Journal of Bone and Mineral Research. 2013;**28**:1243-1255

[7] Demontis GC, Germani MM, Caiaini EG, Barravecchia I, Passino C, Angeloni D. Human pathophysiological adaptations to the space environment. Frontiers in Physiology. 2017;**8**:547. DOI: 10.3389/fphys.2017.00547

[8] Ghosh P, Stabley JN, Behnke BJ, Allen MR, Delp MD. Effects of spaceflight on the murine mandible: Possible factors mediating skeletal changes in non-weight bearing bones of the head. Bone. 2016;**83**:156-161

[9] Zhang B, Cory E, Bhattacharya R, Sah R, Hargens AR. Fifteen days of microgravity causes growth in calvaria of mice. Bone. 2013;**56**:290-295

[10] Hand AR. Salivary glands, salivary secretion, and saliva. In: Hand AR, Frank ME, editors. Fundamentals of Oral Histology and Physiology. Ames: Wiley Blackwell; 2014. pp. 223-240

[11] Pedersen AML, Sørensen CE, Proctor GB, Carpenter GH, Ekström J. Salivary secretion in health and disease. Journal of Oral Rehabilitation. 2018;**45**:730-746

[12] Mednieks M, Khatri A, Rubenstein R, Burleson JA, Hand AR. Microgravity alters the expression of salivary proteins. Oral Health and Dental Management. 2014;**13**:211-216

[13] Mednieks M, Khatri A, Hand AR. Salivary gland protein expression after Bion-M1 and space shuttle STS-135 missions. Gravitational and Space Research. 2015;**3**:2-19

[14] Hand AR. The effects of acute starvation on parotid acinar cells. Ultrastructural and cytochemical observations on ad libitum-fed and starved rats. The American Journal of Anatomy. 1972;**135**:71-92

[15] Hand AR, Ho B. Liquid-diet-induced alterations of rat parotid acinar cells studied by electron microscopy and enzyme cytochemistry. Archives of Oral Biology. 1981;**26**:369-380

[16] Hand AR, Ball WD. Ultrastructural immunocytochemical localization of secretory proteins in autophagic vacuoles of parotid acinar cells of

starved rats. Journal of Oral Pathology. 1988;**17**:279-286

[17] Takahashi S, Uekita H, Kato T, Yuge F, Ushijima N, Inoue K, et al. Involvement of apoptosis and proliferation of acinar cells in atrophy of rat parotid glands induced by liquid diet. Journal of Molecular Histology. 2012;**43**:761-766

[18] Sun GS, Tou JC, Yu D, Girten BE, Cohen J. The past, present, and future of National Aeronautics and Space Administration spaceflight diet in support of microgravity rodent experiments. Nutrition. 2014;**30**:125-130

[19] Hand AR, Weiss RE. Effects of streptozotocin-induced diabetes on the rat parotid gland. Laboratory Investigation. 1984;**51**:429-440

[20] Lotti LV, Hand AR. Endocytosis of parotid salivary proteins by striated duct cells in streptozotocin-diabetic rats. The Anatomical Record. 1988;**221**:802-811

[21] Coleman R, Hand AR. Endocytosis of native and cationized ferritin by intralobular duct cells of the rat parotid gland. Cell and Tissue Research. 1987;**249**:577-586

[22] Hand AR, Coleman R, Mazariegos MR, Lustmann J, Lotti LV. Endocytosis of proteins by salivary gland duct cells. Journal of Dental Research. 1987;**66**:412-419

[23] Lotti LV, Hand AR. Endocytosis of native and glycosylated bovine serum albumin by duct cells of the rat parotid gland. Cell and Tissue Research. 1989;**255**:333-342

[24] Takahashi S, Uekita H, Kato T, Yuge F, Ushijima N, Inoue K, et al. Immunohistochemical and ultrastructural investigation of acinar cells in submandibular and sublingual glands of rats fed a liquid diet. Tissue & Cell. 2014;**46**:136-143

[25] Robinson CP, Bounous DI, Alford CE, Nguyen KH, Nanni JM, Peck AB, et al. PSP expression in murine lacrimal glands and function as a bacteria binding protein in exocrine secretions. American Journal of Physiology. Gastrointestinal and Liver Physiology. 1997;**272**:G863-G871

[26] Geetha C, Venkatesh SG, Dunn BH, Gorr SU. Expression and anti-bacterial activity of human parotid secretory protein (PSP). Biochemical Society Transactions. 2003;**31**:815-818

[27] Bennick A. Salivary proline-rich proteins. Molecular and Cellular Biochemistry. 1982;**45**:83-99

[28] Bennick A. Interaction of plant polyphenols with salivary proteins. Critical Reviews in Oral Biology and Medicine. 2002;**13**:184-196

[29] Mednieks MI, Hand AR. Cyclic AMP binding proteins in saliva. Experientia. 1984;**40**:945-947

[30] Mullins JJ, Mullins LJ, Dunbar DR, Brammar WJ, Gross KW, Morley SD. Identification of a human ortholog of the mouse Dcpp gene locus, encoding a novel member of the CSP-1/Dcpp salivary protein family. Physiological Genomics. 2006;**28**:129-140

[31] Daniel PB, Walker WH, Habener JF. Cyclic AMP signaling and gene regulation. Annual Review of Nutrition. 1998;**18**:353-383

[32] Gold MG, Gonen T, Scott JD. Local cAMP signaling in disease at a glance. Journal of Cell Science. 2013;**126**:4537-4543

[33] Laukaitis CM, Critser ES, Karn RC. Salivary androgen-binding protein (ABP) mediates sexual isolation in *Mus musculus*. Evolution. 1997;**51**:2000-2005

[34] Dagdeviren D, Beallias J, Khan I, Mednieks MI, Hand AR. Response of the mouse sublingual gland to spaceflight. European Journal of Oral Sciences. 2018;**126**:373-381

[35] Mednieks MI, Hand AR. Microheterogeneity of rat parotid gland proteins after chronic treatment with isoproterenol. Journal of Dental Research. 1984;**63**:87-93

[36] Mednieks MI, Epstein PM, Hachisu R, Hand AR, Esquire RG. Cyclic AMP-reactive proteins in human saliva. Archives of Oral Biology. 1994;**39**:869-875

[37] Person P, Eversole LF, Shklar G, Johnson LC, Moss ML. The effects of cosmic particle radiation on pocket mice aboard Apollo XVII: Appendix II. Evaluation of oral, dental, and skeletal tissues. Aviation, Space, and Environmental Medicine. 1975;**46**:634-638

[38] Groza P, Bordeianu A, Cananáu S, Boca A, Petrescu A, Lungu D. The action of simulated and true weightlessness on the digestive tract of rats. Advances in Space Research. 1981;**1**:179-185

[39] Groza P, Bordeianu A, Boca A. Modifications of the digestive tract in rats submitted to an orbital flight aboard the Sovlet satellite Cosmos 1129. Physiologie. 1983;**20**:35-44

[40] Brown LR, Frome WJ, Wheatcroft MG, Riggan LJ, Bussell NE, Johnston DA. The effect of Skylab on the chemical composition of saliva. Journal of Dental Research. 1977;**56**:1137-1143

[41] Groza P, Ursea N, Vasilescu A, Munteanu A, Lungu D, Bolocan N. Changes in some digestive enzymes after a seven-day orbital flight. Physiologie. 1983;**20**:27-33

[42] Johnson P. Saliva: A way to obtain specimens in an operational setting. Aviation, Space, and Environmental Medicine. 1986;**57**:397

[43] Larina IM, Bystritzkaya AF, Smirnova TM. Psycho-physiological monitoring in real and simulated space flight conditions. Journal of Gravitational Physiology. 1997;**4**:P113-P114

[44] Strollo F, Strollo G, Morè M, Ferretti C, Mangrossa N, Casarosa E, et al. Changes in human adrenal and gonadal function onboard SpaceLab. Journal of Gravitational Physiology. 1997;**4**:P103-P104

[45] Buchheim J-I, Matzel S, Rykova M, Vassilieva G, Ponomarev S, Nichiporuk I, et al. Stress related shift toward inflammaging in cosmonauts after long-duration space flight. Frontiers in Physiology. 2019;**10**(85):1-13. DOI: 10.3389/fphys.2019.00085

[46] Bilancio G, Cavallo P, Lombardi C, Guarino E, Cozza V, Giordano F, et al. Urea and minerals monitoring in space missions by spot samples of saliva and urine. Aerospace Medicine and Human Performance. 2019;**90**:43-47

[47] Garrett-Bakelman FE, Darshi M, Green SJ, Gur RC, Lin L, Macias BR, et al. The NASA Twins Study: A multidimensional study of a year-long human spaceflight. Science. 2019;**364**(1-20):144. DOI: 10.1126/science.aau8650

[48] Dagdeviren D, Kalajzic Z, Adams DJ, Kalajzic I, Lurie A, Mednieks MI, et al. Responses to spaceflight of mouse mandibular bone and teeth. Archives of Oral Biology. 2018;**93**:163-176

[49] Andreev-Andrievskiy A, Popova A, Boyle R, Alberts J, Shenkman B, Vinogradova O, et al. Mice in Bion-M 1 space mission: Training and selection. PLoS One. 2014;**9**:e104830

[50] Novoselova EG, Lunin SM, Khrenov MO, Parfenyuk SB, Novoselova TV, Shenkman BS, et al. Changes in immune cell signalling, apoptosis and stress response functions in mice returned from the BION-M1 mission in space. Immunobiology. 2015;**220**:500-509

[51] Bresin A. Effects of masticatory muscle function and bite-raising on mandibular morphology in the growing rat. Swedish Dental Journal. Supplement. 2001;**150**:1-49

[52] Kiliaridis S. Masticatory muscle function and craniofacial morphology. An experimental study in the growing rat fed a soft diet. Swedish Dental Journal. Supplement. 1986;**36**:1-55

[53] Beamer WG, Donahue LR, Rosen CJ, Baylink DJ. Genetic variability in adult bone density among inbred strains of mice. Bone. 1996;**18**:397-403

[54] Somerville JM, Aspden RM, Armour KE, Armour KJ, Reid DM. Growth of C57Bl/6 mice and the material and mechanical properties of cortical bone from the tibia. Calcified Tissue International. 2004;**74**:469-475

[55] Lightfoot JT. Sex hormones' regulation of rodent physical activity: A review. International Journal of Biological Sciences. 2008;**4**:126-132

[56] Widmer CG, Morris-Wiman J. Assessment of incising ethology in the absence and presence of jaw muscle hyperalgesia in a mouse home cage. Physiology & Behavior. 2015;**149**:229-238

[57] Simmons DJ, Russell JE, Winter F, Tran Van P, Vignery A, Baron R, et al. Effect of spaceflight on the non-weight-bearing bones of rat skeleton. American Journal of Physiology. Regulatory, Integrative and Comparative Physiology. 1983;**244**:R319-R326

[58] Simmons DJ, Grynpas MD, Rosenberg GD. Maturation of bone and dentin matrices in rats flown on the Soviet biosatellite Cosmos 1887. The FASEB Journal. 1990;**4**:29-33

[59] Morey-Holton E, Globus RK, Kaplansky A, Durnova G. The hindlimb unloading rat model: Literature overview, technique update and comparison with space flight data. Advances in Space Biology and Medicine. 2005;**10**:7-40

[60] Globus RK, Morey-Holton E. Hindlimb unloading: Rodent analog for microgravity. Journal of Applied Physiology (Bethesda, MD: 1985). 2016;**120**:1196-1206

[61] Simmons DJ, Grazman B, Russell JE, Walker WV, Bikle DD, Morey ER. Simulating certain aspects of hypogravity: Effects on bone maturation in the non-weight bearing skeleton. Aviation, Space, and Environmental Medicine. 1983;**54**:1080-1084

[62] Globus RK, Bikle DD, Morey-Holton E. Effects of simulated weightlessness on bone mineral metabolism. Endocrinology. 1984;**114**:2264-2270

[63] Guo R, Hu M, Sun ZY. Xue JW. [Effects of simulated weightlessness on rats mandible, lumbar vertebra and femur]. Space Medicine & Medical Engineering. 2005;**18**:165-169

[64] Roer RD, Dillaman RM. Bone growth and calcium balance during simulated weightlessness in the rat. Journal of Applied Physiology (Bethesda, MD: 1985). 1990;**68**:13-20

[65] Winkler DG, Sutherland MK, Geoghegan JC, Yu C, Hayes T, Skonier JE, et al. Osteocyte control of bone formation via sclerostin, a novel BMP antagonist. The EMBO Journal. 2003;**22**:6267-6276

[66] Weivoda MM, Oursler MJ. Developments in sclerostin biology: Regulation of gene expression, mechanisms of action, and physiological functions. Current Osteoporosis Reports. 2014;**12**:107-114

[67] Robling AG, Niziolek PJ, Baldridge LA, Condon KW, Allen MR, Alam I, et al. Mechanical stimulation of bone in vivo reduces osteocyte expression of Sost/sclerostin. The Journal of Biological Chemistry. 2008;**283**:5866-5875

[68] Spatz JM, Fields EE, Yu EW, Divieti Pajevic P, Bouxsein ML, Sibonga JD, et al. Serum sclerostin increases in healthy adult men during bed rest. The Journal of Clinical Endocrinology and Metabolism. 2012;**97**:E1736-E1740

[69] Macias B, Swift JM, Nilsson MI, Hogan HA, Bouse SD, Bloomfield SA. Simulated resistance training, but not alendronate, increases cortical bone formation and suppresses sclerostin during disuse. Journal of Applied Physiology. (Bethesda, MD: 1985). 2012;**112**:918-925

[70] Macias BR, Lima F, Swift JM, Shirazi-Fard Y, Greene ES, Allen MR, et al. Simulating the lunar environment: Partial weightbearing and high-LET radiation-induce bone loss and increase sclerostin-positive osteocytes. Radiation Research. 2016;**186**:254-263

[71] Smith SM, Heer M, Shackelford LC, Sibonga JD, Spatz J, Pietrzyk RA, et al. Bone metabolism and renal stone risk during International Space Station missions. Bone. 2015;**81**:712-720

[72] Vico L, van Rietbergen B, Vilayphiou N, Linossier M-T, Locrelle H, Normand M, et al. Cortical and trabecular bone microstructure did not recover at weight-bearing skeletal sites and progressively deteriorated at non-weight-bearing sites during the year following International Space Station missions. Journal of Bone and Mineral Research. 2017;**32**:2010-2021

[73] Blaber EA, Dvorochkin N, Lee C, Alwood JS, Yousuf R, Pianetta P, et al. Microgravity induces pelvic bone loss through osteoclastic activity, osteocytic osteolysis, and osteoblastic cell cycle inhibition by CDKN1a/p21. PLoS One. 2013;**8**(4):e61372. DOI: 10.1371/journal.pone.0061372

[74] Gerbaix M, Gnyubkin V, Farlay D, Olivier C, Ammann P, Courbon G, et al. One-month spaceflight compromises the bone microstructure, tissue-level mechanical properties, osteocyte survival and lacunae volume in mature mice skeletons. Scientific Reports. 2017;**7**:2659. DOI: 10.1038/s41598-017-03014-2

[75] Hauschka PV, Lian JB, Cole DEC, Gundberg CM. Osteocalcin and matrix gla protein: Vitamin K-dependent proteins in bone. Physiological Reviews. 1989;**69**:990-1047

[76] Yang C, Chen J, Wu F, Li J, Liang P, Zhang H, et al. Effects of 60-day headdown bed rest on osteocalcin, glycolipid metabolism and their association with or without resistance training. Clinical Endocrinology. 2014;**81**:671-678

[77] Karsenty G. Update on the biology of osteocalcin. Endocrine Practice. 2017;**23**:1270-1274

[78] Moser SC, van der Eerden BCJ. Osteocalcin—A versatile bone-derived hormone. Frontiers in Endocrinology. 2019;**9**:794. DOI: 10.3389/fendo.2018.00794

[79] Patterson-Buckendahl P, Arnaud SB, Mechanic GL, Martin RB, Grindeland RE, Cann CE. Fragility and composition of growing rat bone after one week in spaceflight. American Journal of Physiology. Regulatory, Integrative and Comparative Physiology. 1987;**252**:R240-R246

[80] Backup P, Westerlind K, Harris S, Spelsberg T, Kline B, Turner R. Spaceflight results in reduced mRNA levels for tissue-specific proteins in the musculoskeletal system. American Journal of Physiology. Endocrinology and Metabolism. 1994;**266**:E567-E573

[81] Bikle DD, Harris J, Halloran BP, Morey-Holton E. Altered skeletal pattern of gene expression in response to spaceflight and hindlimb elevation. American Journal of Physiology. Endocrinology and Metabolism. 1994;**267**:E822-E827

[82] Evans GL, Morey-Holton E, Turner RT. Spaceflight has compartment- and gene-specific effects on mRNA levels for bone matrix proteins in rat femur. Journal of Applied Physiology. (Bethesda, MD: 1985). 1998;**84**:2132-2137

[83] Collet P, Ueebelhart D, Vico L, Moro L, Hartmann D, Roth M, et al. Effects of 1 and 6 months spaceflight on bone mass and biochemistry in two humans. Bone. 1997;**20**:547-551

[84] Caillot-Augusseau A, Lafage Proust MH, Soler C, Pernod J, Dubois F, Alexandre A. Bone formation and resorption biological markers in cosmonauts during and after a 180-day space flight (Euromir 95). Clinical Chemistry. 1998;**44**:578-585

[85] Caillot-Augusseau A, Vico L, Heer M, Voroviev D, Souberbielle JC, Zitterman A, et al. Space flight is associated with rapid decreases of undercarboxylated osteocalcin and increases of markers of bone resorption without changes in their circadian variation: Observations in two cosmonauts. Clinical Chemistry. 2000;**46**:1136-1143

[86] Carmeliet G, Nys G, Bouillon R. Microgravity reduces the differentiation of human osteoblastic MG-63 cells. Journal of Bone and Mineral Research. 1997;**12**:786-794

[87] Harris SA, Zhang M, Kidder LS, Evans GL, Spelsberg TC, Turner RT. Effects of orbital spaceflight on human osteoblastic cell physiology and gene expression. Bone. 2000;**26**:325-331

[88] Kapitonova MY, Kuznetsov SL, Salim N, Othman S, Kamauzaman TM, Ali AM, et al. Morphological and phenotypical characteristics of human osteoblasts after short-term space mission. Bulletin of Experimental Biology and Medicine. 2014;**156**:393-398

[89] Catón J, Tucker AS. Current knowledge of tooth development: Patterning and mineralization of the murine dentition. Journal of Anatomy. 2009;**214**:502-515

[90] Kiliaridis S. The relationship between masticatory function and craniofacial morphology. III. The eruption pattern of the incisors in the growing rat fed a soft diet. European Journal of Orthodontics. 1986;**8**:71-79

[91] Burn-Murdoch RA. The effect of the consistency of the diet on eruption rates and lengths of incisor teeth in rats. Archives of Oral Biology. 1993;**38**:699-706

[92] Soenjaya Y, Foster BL, Nociti FH Jr, Ao M, Holdsworth D, Hunter GK, et al. Mechanical forces exacerbate periodontal defects in Bsp-null mice. Journal of Dental Research. 2015;**94**:1276-1285

[93] Prokhonchukov AA, Tigranian RA, Kolesnik AG, Novikov LL, Timofeeva NT. [Calcium and phosphorus content and the 45Ca incorporation into the bones and teeth of rats after a 22-day orbital space flight on board the "Cosmos-605" satellite ship]. Kosmicheskaia Biologiia i Aviakosmicheskaia Meditsina. 1977;**11**:26-30

[94] Simmons DJ, Russell JE, Winter F, Baron R, Vignery A, Tran VT, et al. Bone growth in the rat mandible during space flight. The Physiologist. 1980; **23**(Suppl 6):S87-S90

[95] Rosenberg GD, Campbell SC, Simmons DJ. The effects of spaceflight on the mineralization of rat incisor dentin. Proceedings of the Society for Experimental Biology and Medicine. 1984;**175**:429-437

[96] Rosenberg GD, Simmons DJ. Electron microprobe analyses of Ca, S, Mg and P distribution in incisors of Spacelab-3 rats. The Physiologist. 1985;**28**(6 Suppl):S189-S190

[97] Prokhonchukov AA, Komissarova NA, Zhizhina NA, Volozhin AI. [Comparative study of the effect of weightlessness and artificial gravity on the density, ash, calcium, and phosphorus content of calcified tissues]. Kosmicheskaia Biologiia i Aviakosmicheskaia Meditsina. 1980;**14**:23-26

[98] Tang C, Niu Z, Zheng Y, Chen Y, Bao B. Meng Q. [Effects of hypergravity exposure after 30-days simulated weightlessness on chemokine CCL20 and its receptor CCR6 in lingual mucosa of rhesus macaque]. Zhonghua Yi Xue Za Zhi. 2014;**94**:2525-2530

[99] Philippou A, Minozzo FC, Spinazzola JM, Smith LR, Lei H, Rassier DE, et al. Masticatory muscles of mouse do not undergo atrophy in space. The FASEB Journal. 2015;**29**:2769-2779

[100] Mukaibo T, Gao X, Yang NY, Oei MS, Nakamoto T, Melvin JE. Sexual dimorphisms in the transcriptomes of murine salivary glands. FEBS Open Bio. 2019;**9**:947-958. DOI: 10.1002/2211-5463.12625

[101] Michael D, Soi S, Cabera-Perez J, Weller M, Alexander S, Alevizos I, et al. Microarray analysis of sexually dimorphic gene expression in human minor salivary glands. Oral Diseases. 2011;**17**:653-661

Chapter 5

Space Radiation-Induced Hematopoietic Stem Cell Injury

Huihong Zeng, Mengzhen Yue and Lijian Shao

Abstract

Space radiation is an unavoidable health risk during space activities. Hematopoietic cells are sensitive to radiation including proton and oxygen radiation and so on. Understanding the mechanisms responsible for detrimental effects of space radiation is important to achieve countermeasures protecting hematopoietic stem cells (HSCs), which generates different hematopoietic populations. However, the biological effects of various sources of space radiation on HSCs are not understood well. Induction of cellular apoptosis, reactive oxygen species (ROS), and DNA damage upon space radiation is believed to be critical mediators for HSC damage. In this chapter, we will mainly discuss the biological effectiveness of proton and oxygen radiation on the numbers and function of HSCs. Space radiation-induced apoptosis, ROS, and DNA damage were examined as well, which will provide foundation to develop novel strategies protecting HSCs from space radiation.

Keywords: space irradiation, proton irradiation, oxygen irradiation, apoptosis, reactive oxygen species, hematopoietic stem cells, bone marrow

1. Introduction

Human spaceflight and exploration began in the 1960s. Manned spaceflight activities have continually expanded in frequency and scope since that time, and plans are now forming for long-duration flights to deep-space destinations. However, numerous risk factors have potential to negatively affect the astronauts' health during deep-space missions, especially microgravity and space radiation. Exposure of astronauts to space radiation is relatively unpredictable yet inevitable. Space radiation comes from two major sources: solar particle events (SPE) emanating from the sun and galactic cosmic rays (GCR) originating from sources outside the solar system.

SPE mainly includes protons and can lead to moderate- to high-dose rate exposures to ionizing radiation during long-term space mission [1, 2]. Astronauts may receive cumulative doses from 1 to 3 gray (Gy) during an SPE [3, 4]. Especially, proton radiation contributes to more than 80% of SPE [1, 2, 5].

GCR contains high atomic number and energy (HZE) particles, such as ^{56}Fe, ^{28}Si, ^{16}O, ^{12}C, and so forth. HZE particles are characterized by dense tracks of ionization, a property quantified as high-linear energy transfer (LET). The properties of HZE particles are consistent with their stronger toxicities and higher energy to normal tissues than photon and proton radiation [6, 7]. Previous studies have documented that the value of RBE in relation to γ-ray radiation was 1.25 for ^{56}Fe, 1.4 for ^{28}Si, and 0.99 for ^{12}C using a mouse model [6, 8]. Among HZE particles, ^{56}Fe has high-linear energy transfer that might heavily contribute to GCR in space [9]. There are many different components including ions, hydrogen, helium, and so on in spacecraft [10]. Based on measurements in the Mars Science Laboratory from 2011 to 2012, the irradiation dose of GCR in the spaceflight was approximately 481 ± 80 μGy per day [11]. For a 600- to 900-day Mars mission, the total radiation doses from GCR reside between 0.33 and 0.49 Gy. Therefore, the total radiation dose from SPE and GCR will reach to 1.0 Gy or above. Although doses and dose rates of space radiation are low, it will still result in space dose accumulation in the body and high risk to astronauts' health during a long-term space mission [9, 12].

Radiation-induced tissue damage in the body has long been understood since Wilhelm Röntgen discovered X-ray in 1895 [13]. Hematopoietic and gastrointestinal systems have been shown to be the two most sensitive compartments of the body to radiation. It has been well-documented that radiation (including space radiation) also induces dysfunction of the brain, manifesting as behavioral and cognitive disabilities [14, 15]. The detrimental effects of X-ray radiation in the body were firstly reported by Warren and Whipple [16] and Shouse et al. [13]. They reported that exposing dogs to high doses of X-rays resulted in death from severe hematopoietic suppression and damage. The detrimental effects of radiation on human health were heavily realized after the use of the first atomic bombs in 1945. Many people in Hiroshima and Nagasaki who survived the initial bomb blast later died from radiation exposure in the event. Long-term toxic impacts of the atomic bomb on humans were observed as well, such as the high risk of hematopoietic malignancies. Further studies proved that hematopoietic failure was one of the primary reasons in radiation-induced death when animals experienced a moderate to high dose of total-body irradiation. This is supported by a study in the 1940s showing that shielding the spleen or one entire hind leg with lead or transplantation of splenocytes protected mice from the lethal effect of irradiation [17]. The importance of hematopoietic cells under radiation was also supported by studies showing that intravenous infusions of bone marrow (BM) cell suspensions protected mice from the effects of radiation [95]. Initially, investigators suggested that a humoral factor from the spleen and BM cell suspensions might benefit the radioprotective effects [18], while later studies proved that it was attributed to the transplanted hematopoietic cells [19–21]. When Till and McCulloch discovered hematopoietic stem cells (HSCs) in the 1960s, those cells protecting animals from IR-induced lethal hematopoietic damage were HSCs [22, 23]. Remarkable progress has been subsequently made in understanding of the mechanisms by which radiation causes hematopoietic damage.

However, the effects of space radiation on the hematopoietic system have yet to be fully understood, leading to a lack of effective countermeasure strategies thus far. In the present chapter, we mainly focus our discussion on the biological effectiveness of space radiation, such as proton and oxygen, whereby space radiation induces HSC injury, and the implication of HSC injury to IR-induced BM suppression in mouse. In addition, genomic instability, malignancies, and intestinal, brain, behavioral, and cognitive effects induced by space radiation will not be discussed here, which were extensively discussed by other investigators [24, 25].

2. Composition of the mouse hematopoietic system

Hematopoietic stem cells often replenish the whole blood system throughout the life span of the body and maintain hematopoietic homeostasis. In Till and McCulloch's landmark work [22, 23, 26], they utilized colony-forming unit-spleen (CFU-S) assay to identify functional HSCs even while these HSCs were a mixed population with different proliferating and self-renewal abilities. This pioneering research not only provided strong evidence to show in vivo HSC existence but also stimulated many investigators to define and isolate HSCs and characterize their properties in mice and humans.

In the decades since Till and McCulloch's study [22, 23, 26], multiple cell surface markers have been used to separate HSCs from other populations. For example, lineage cell surface markers, such as Ter119, CD4, CD8, B220, Gr-1, Mac-1, and NK1.1, are not detected neither in mouse multipotent progenitors (MPPs) nor in hematopoietic progenitor cells (HPCs). Both HSCs and MPPs express c-Kit and Scal-1, which are therefore named lineage negative (Lin-), c-Kit$^+$, and Sca-1$^+$ as LSK cells (**Figure 1**). HPCs express c-Kit, but not Sca-1, and are termed Lin$^-$c-Kit$^+$Sca-1$^-$. In 2005, Kiel et al. further utilized CD150 and CD48 surface markers to distinguish HSCs and MPPs [27]. HSCs express CD150, but not CD48, and termed CD150$^+$CD48$^-$LSK cells. MPPs are CD150$^{+/-}$CD48$^+$LSK cells. Investigators also used different strategies to identify HSCs. Combination of CD34 or Thy^{lo}Flk-2$^-$ with LSK surface markers was used to isolate HSCs, named as CD34$^-$LSK and Thy1^{lo}Flk-2$^-$LSK cells, respectively [26]. Trumpp's group combined CD34, CD135, CD150, CD48, and LSK markers to further differentiate HSC into long-term HSCs (CD34$^-$CD135$^-$CD150$^+$CD48$^-$LSK cells) and short-term HSCs (CD34$^+$CD135$^-$CD150$^+$CD48$^-$LSK cells). In addition, HSCs have a feature with highly expression of ATP-binding protein like ABCG2 and can efflux DNA-binding dye Hoechst 33342. The Hoechst-effluxing side population (SP) cells were therefore used to indicate HSC population in the case of some circumstances [28], such as 5FU and radiotherapeutic stress conditions.

It is well-documented that the hematopoietic system is organized in a hierarchical manner (**Figure 1**). The rare long-term HSCs are located at the top of the hierarchy and have the capacity to self-renew, proliferate, and differentiate into various lineages of mature blood cells though immature MPPs and HPCs [29]. The most important feature of HSCs is quiescent to maintain their self-renewal ability

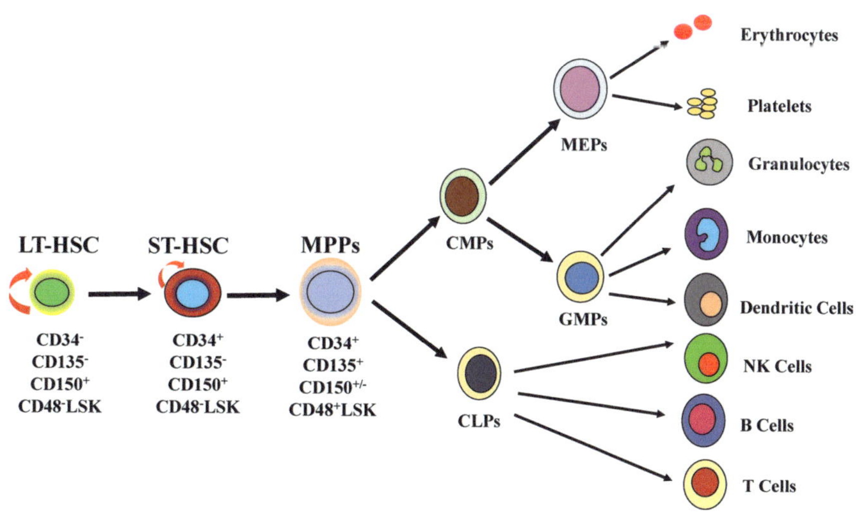

Figure 1.
Composition of mouse hematopoietic system. Long-term hematopoietic stem cell (LT-HSCs, CD34$^-$CD135$^-$CD150$^+$CD48$^-$LSK cells) is located at the top of the hierarchy and has the ability to self-renew, proliferate, and differentiate into short-term HSCs (ST-HSC, CD34$^+$CD135$^-$CD150$^+$CD48$^-$LSK cells) and multipotent progenitors (MPPs, CD34$^+$CD135$^+$CD150$^{+/-}$CD48$^+$LSK cells), which can further differentiate into different progenitors, such as common lymphoid progenitors (CLPs), common myeloid progenitors (CMPs) and their progeny megakaryocyte/erythroid progenitors (MEPs), and granulocyte/monocyte progenitors (GMPs). Progenitors can proliferate different lineages of mature blood cells.

and potentially provide lifelong hematopoiesis. The dominant HSCs have ability to protect the whole blood system against different stress conditions [30]. Under sublethal irradiation, HSCs have been long-term damaged, which can be easily ignored in clinic because of normal cell counts from the bone marrow and peripheral blood. Damaged HSCs induced by photon-irradiation have impaired self-renewing ability, leading to bone marrow failure and death [31]. However, whether low doses of space radiation trigger long-term HSC damage remains unknown. Comparing to HSCs, MPPs and HPCs have limit or lack self-renewal ability even though they are proliferating populations. The property of MPPs and HPCs with proliferation provides a beneficial role in case of normal hematopoiesis and stress hematopoiesis. For example, in case of blood loss or infection, MPP and HPC quickly proliferate to meet the requirement of mature cell production, trying to maintain normal hematopoiesis. Under radiotherapy and chemotherapy, MPPs and HPCs can be easily depleted with acute myelosuppression because of their proliferating feature. This will lead to HSC activation, proliferation, and differentiation to reestablish MPP and HPC populations and rebuild hematopoiesis, which might result in HSC exhaustion.

3. Proton radiation and hematopoietic stem cells

During spaceflight missions outside low Earth orbit, there remains the possibility of astronauts receiving damaging doses of space radiation. Because the dose-depth distribution of SPE spectrum protons is relatively low [32], during an SPE the skin and organs near the surface of the skin will receive higher doses than deeper organs such as the bone marrow. In addition, compared to traditional radiation therapy with X-rays (photons) or electrons, proton therapy has potential benefit for clinical cancer treatment because of its favorable distribution of the radiation dose, leading to selectively increased radiation dose to the cancerous tissues while lowering the dose to normal tissues [33, 34]. However, the hematopoietic system is highly sensitive to ionizing radiation, and exposure to even a relatively low dose of SPE may still be able to result in substantial damage to the system [35, 36]. Therefore, understanding the biological effects of proton radiation is immediately needed.

One of the characteristics of the radiation-induced hematopoietic syndrome is a decline in blood cell counts, resulting from radiation-induced cell killing in circulating blood cells and suppression of hematopoietic stem and progenitor cells in the bone marrow [37–39]. Studies have reported that whole-body exposure to protons causes acute effects on the hematopoietic system in animal models [40–45]. The decreased WBCs, lymphocytes, and neutrophils were detected starting at 4 hours after 0.25–3 Gy proton radiation, with the lowest numbers observed on day 4 [43, 44]. The reduction in WBCs and lymphocytes was still evident in mice after exposure to 2.0 Gy of protons (230 MeV) [41]. This might be due to the high sensitivity of lymphocytes to proton radiation, which is consistent with the data from γ-radiation [46]. Two Gy of proton radiation induced the acute decrement of peripheral blood cells, which was shown completely recovered in the long term. The dynamic changes of peripheral blood counts might result in ignoring the negative effects of proton radiation on hematopoietic system. However, the abnormalities of splenic WBCs and lymphocytes were still detected at more than 100 days after low dose of proton radiation [47]. Taken together, proton radiation has not only acute injury but also long-term harmful effects on the hematopoietic system.

It has been well established that exposure to a significant dose of total-body γ-irradiation (TBI) induces not only the acute radiation hematopoietic syndrome but also long-term bone marrow injury [48, 49]. The acute radiation hematopoietic syndrome induced by γ-irradiation is primarily attributed to the induction of apoptosis of HPCs, while γ-irradiation-induced long-term BM suppression is mainly ascribed to the persistent damage to HSCs. While the effects of γ-irradiation on the hematopoietic system have been extensively documented, much less is known about the effects of proton irradiation [35, 36].

Previously, mice were exposed to two different doses (0.5 and 1.0 Gy) of proton radiation and examined the acute and long-term effects on BM HSPCs at 2 and 22 weeks after proton radiation, respectively (**Table 1**). Results showed that exposure of mice to 1.0 Gy of proton radiation resulted in a significant decrease in the number of WBCs and PLTs from peripheral blood 2 weeks after the exposure [50]. It was demonstrated that 1.0 Gy of oxygen ion radiation (^{16}O TBI) significantly decreased the cell counts of peripheral blood leukocytes when measured 2 weeks after exposure in male C57BL/6 mice [51]. Interestingly, the decrease of peripheral blood cell counts was not observed 2 weeks after 0.5 Gy proton TBI or 0.1 and 0.25 Gy ^{16}O TBI [51]. The threshold dose of protons (50 or 70 MeV) to induce a decline in WBC counts in female ICR outbred mice was previously estimated to be between 0.25 and 0.5 Gy [43]. The threshold dose identified from this previous study is lower than the result from our study. It may have resulted in part from the use of different strains of mice, the time when the mice were studied after radiation or the difference in energies of protons between the two studies. For example, the linear energy transfer values for 50, 70, and 150 MeV protons would be 1.26, 0.96, and 0.55 keV/μm [52]. Because of the dose-depth distribution of SPE protons, only relatively large or higher-energy spectrum SPEs may lead to BM exposure to these doses of protons in astronauts. However, the doses of SPEs that can cause significant HSPC damage have been observed, raising the possibility that astronauts might experience reductions in circulating blood cell counts and BM HSPC damage if they encounter such an SPE. The dose-depth distribution in mice exposed to protons is different from that in humans [1]. It was found that a dose of 0.5 Gy protons (150 MeV) significantly reduced hematopoietic stem/progenitor cell function. The effects of protons at this energy and at doses below 0.5 Gy are unknown. Since the dose to the blood-forming tissues of human subjects will likely be low, future studies need to examine the hematopoietic effects of proton doses below 0.5 Gy.

Proton TBI can acutely induce the decrease of all lineages of peripheral blood cells, since all lineages of blood cells are generated from hematopoietic stem cells through their differentiation into various lineages of progenitors. We have shown that the exposure to both 0.5 and 1.0 Gy proton TBI damaged not only HPCs but also LSK cells, leading to the defect in their numbers and function, which were supported by the decreased abilities to form in vitro colonies including BFU-E,

	Number of cells		Function of HSCs		
	HPC	LSK cells	BFU-E	CFU-GM	CFU-GEMM
2 weeks	↓	↓	↓	↓	↓
22 weeks	no change	↓	↓	↓	↓

Note: C57BL/6 mice were exposed to 1.0 Gy of proton total body irradiation. The indicated parameters were measured at 2 and 22 weeks after the exposure.
↓ Indicate that the parameter is decreased when compared to non-irradiated mice.
No change means that the parameter is similar to that from non-irradiated mice.

Table 1.
Acute and long-term effects of proton irradiation on hematopoietic cells.

CFU-GM, and CFU-GEMM [50]. The in vivo functional defect of hematopoietic stem and progenitor cells after proton exposure will be further investigated through bone marrow transplantation in future studies. We have also reported that a 1.0 Gy dose of ^{16}O TBI significantly decreased peripheral blood counts and BM HSPCs 2 weeks after the exposure [51]. Therefore, the mechanisms of space radiation-induced acute damage to the hematopoietic system should be investigated further.

In our previous studies, we firstly show that exposure to proton radiation causes long-term hematopoietic injury at 22 weeks after the exposure [53]. Our data provide the first direct evidence that exposure of mice to 1.0 Gy dose of proton radiation results in not only a sustained reduction in the frequency of BM HSCs but also in the long-term inhibition of HSCs clonogenic function to form BFU-E, CFU-GM, and CFU-GEMM colonies in vitro (**Table 2**). In contrast, the number and frequency of HPCs returned to normal levels at 22 weeks postradiation. Another question that needs to be addressed is whether proton radiation-induced HSC damage leads to hypoplastic syndrome after hematopoietic stress. Myeloid leukemia could be induced by low and/or moderate doses of γ-irradiation and ^{56}Fe heavy ion radiation in mice [54, 55]. It has been shown that proton radiation induced minor myeloid leukemia but have high possibility to induce hepatocellular adenoma and malignant lymphoma in CBA mice [56], which is in contrast with the effects of γ-irradiation, showing that 3 Gy of γ-irradiation caused around 25% of CBA mice developing acute myeloid leukemia. These differences are due to the differential biological effectiveness between proton and γ-irradiation. To closely mimic space environment, investigators used minipig animal models to expose to electron solar particle event (eSPE) [57]. Comparing to eSPE, proton solar particle events (pSPE) have stronger negative effect on the numbers of peripheral WBCs, lymphocytes, neutrophils, and monocytes with a factor of 2.79. These data suggest that different hematopoietic populations have differential radiosensitivity to proton irradiation.

Protons have a higher linear energy transfer and are denser ionizing radiation than photon. Therefore, protons deposit high energy at the end of their range termed "Bragg peak" and cause heavily damage to the target tissues, cells, and molecules. Compared to γ-irradiation, proton radiation causes larger γ-H2AX foci [58], leads to hypermethylated DNA [59], has different transcriptome profiles [60], and modulates different signaling pathways [61]. More detailed investigation into what unique biological effects proton radiation has is called for to instruct proton studies.

Because an SPE contains protons of multiple energies below 150 MeV, some groups have developed cell culture or animal models of exposure to broad energy spectra of protons to better simulate an SPE. Previously, some differences were

	HPC			LSK cells		
	Apoptosis	ROS	DNA damage	Apoptosis	ROS	DNA damage
2 weeks	↑	no change	no change	↑	↑	no change
22 weeks	no change	no change	no change	no change	↑	↑

Note: C57BL/6 mice were exposed to 1.0 Gy of proton total body irradiation. The indicated parameters were measured at 2 and 22 weeks after the exposure.
↓ or ↑ Indicate that the parameter is decreased or increased when compared to non-irradiated mice, respectively. No change means that the parameter is similar to that from non-irradiated mice.

Table 2.
Induction of cellular apoptosis, ROS, and DNA damage after proton irradiation.

found in peripheral blood cell counts between mice exposed to one-energy protons (230 MeV, 2 Gy) and mice exposed to SPE-like protons at the same dose level [41]. Exposure of minipigs to SPE-like protons at a skin dose as low as 0.5 Gy (estimated dose to the BM: 0.42 Gy) caused decrement in peripheral blood cell counts up to at least 2 weeks after exposure [45]. Hence, further studies with SPE-like proton exposures at low doses are warranted. Since we used only two radiation doses (0.5 and 1 Gy), we were not able to prepare dose-response curves for the effects observed. In one of our separate studies, mice were irradiated with fully modulated beams containing particles from 0 energy to 150 MeV and a uniform dose versus depth profile. Doses of protons were 0.1, 0.25, and 0.5 Gy. Bone marrow cells were collected at 2 weeks after irradiation and examined as described in the current manuscript. We found a dose-dependent decrease in LSK cells, together with an increase in ROS levels and apoptosis in these cells (data not shown). In summary, our study has demonstrated that acute exposures to medium doses of proton TBI induced damage to HPCs and LSK cells in a mouse model.

Radiation-induced cell damage might be mediated by induction of apoptosis, DNA damage, and oxidative stress [48, 62]. We therefore assessed those parameters in HPCs and LSK cells 2 weeks after proton exposure [50]. Our data indicate that HPCs and LSK cells may respond differently to proton radiation. Exposure to 1.0 Gy protons resulted in an increase in cellular apoptosis in HPCs. Irradiated LSK cells, on the other hand, showed both increased apoptosis and oxidative stress. Neither of the two cell types showed enhanced DNA damage or cell cycling 2 weeks after proton exposure. Importantly, LSK cells in mice bone marrow from acute and long-term proton exposure cause the significant induction of oxidative stress [50, 53]. In the previous studies, it has been reported that the long-term increase in ROS production in LSK cells was observed after ^{16}O radiation and γ-rays [48, 63, 64]. For example, increasing levels of ROS production were detected at 2 months after 6.0 Gy of total-body γ-irradiation, which might be related to irradiation-induced DNA damage, leukemia, and senescence in irradiated hematopoietic stem and progenitor cells. Both proton and gamma radiation may induce residual negative effects on the bone marrow, which might be mediated by overproduction of chronic reactive oxidative stress in HSCs [63, 65, 66]. Taken together, present data indicate that irradiation-induced oxidative stress in HSCs might be a critical factor in the hematopoietic cell response to space radiation.

Reactive oxygen species (ROS) plays an important role in determining the fate of normal stem cells. Low levels of ROS are required for stem cells to maintain their quiescence and self-renewal capacities. Increases in ROS production cause stem cell proliferation, differentiation, apoptosis, and cell death, leading to their exhaustion. Regulating ROS production in stem cells is important to maintain tissue homeostasis and repair damaged area during the life span of an organism. It has been reported that the levels of ROS were closely related to the proper functional hematopoietic stem cells. There are multiple different ways for ROS production in cells, such as mitochondria oxidative phosphorylation, glycolysis, NADPH oxidases (NOXs) enzyme, peroxisomal and cytochrome P450 metabolism, and so on. Mitochondria oxidative phosphorylation is not a major source to generate ROS in hematopoietic stem cells under homeostasis. This is because (1) HSCs locate bone marrow hypoxic niche with low levels of oxygen; (2) HSCs have small amount of mitochondrial and immature mitochondrial; (3) HSCs have high level of pimonidazole, which is a hypoxia marker; and (4) HSCs have capacity to response to hypoxia by increasing hypoxia-inducible factor 1α (HIF-1α) expression [67]. Subsequently, increasing levels of HIF-1α benefit HSCs to use anaerobic glycolysis, instead of mitochondrial oxidative phosphorylation, to produce energy along with reducing ROS production. Previous studies have shown that increasing expression of NOX enzyme might contribute to γ-irradiation-induced ROS production in HSCs [66], which was supported by increasing NOX4 expression, rather than other isoforms of NOXs in γ-ray irradiated HSCs [66]. Diphenyliodonium (a selective NOX inhibitor) treatment can partially restore the functional impairment in γ-ray irradiated HSCs by decreasing irradiation-induced ROS production in HSCs [66]. For space radiation circumstance, the higher level of NOX4 expression was observed in proton-irradiated HSCs than that in unirradiated HSCs. These data indicated that NOX enzyme, especially NOX4, might be involved in the induction of ROS production in proton-irradiated HSCs. The importance of ROS production in space radiation-induced HSC injury should further be assessed by using NOX4 inhibitor or other antioxidants, such as N-acetylcysteine and gamma-tocotrienol in future studies. There are some other potential unanswered questions including (1) whether these chronically oxidative-stressed HSCs induced by proton radiation experience senescence, (2) whether these space-irradiated HSCs have chromosomal instability, and (3) whether the chromosomal aberrant HSCs after space radiation result in the leukemia development, which was evidenced in mice after γ-irradiation exposure [38, 65, 68].

One of HSC properties is its self-renewal ability, which is sustained via its slow cycling and quiescence. By using BrdU-chasing assay and H2B-GFP mice model, it has been shown that dominant HSCs divide only once every 145 days, which ensures self-renewal capacity along with providing whole life blood homeostasis and avoiding HSC exhaustion [69]. It was reported that loss of FOXO3a resulted in increasing ROS production and accelerating HSC cycling, which is along with the defect of HSC self-renewal capacity and the exhaustion of HSCs. N-acetylcysteine (NAC), an antioxidant, can protect FOXO3a mutant HSCs from oxidative stress and restore HSC dormancy. The same phenotype was also seen in the case of loss of Bmi-1 and TSC1 in mice [70–72], which is due to increasing ROS production and cycling in HSCs. Upon proton radiation, data have shown that there were far fewer HSCs in G0 phase and higher numbers of HSCs in G_1 phase than nonirradiated controls. Proton radiation-induced HSC cycling is consistent with upregulation of positive cell cycle regulators cyclin D1 and cyclin D3. Although HSC proliferation might compensate for the decreased number of HSCs after proton radiation, it will be at risk of loss self-renewal of HSCs.

Additionally, it has demonstrated that the persistent increase of DNA damage in proton-irradiated HSCs, but not in HPCs, was associated with proton radiation-induced ROS production in HSCs. Unrepaired DNA damage in proton-irradiated HSCs might negatively affect HSC self-renewal, proliferation, and differentiation, leading to long-term functional damage in HSCs. Taken together, these findings provide strong evidence showing that proton TBI induces not only acute hematopoietic injury but also long-term BM suppression and HSC damage. These detrimental effects of proton radiation on hematopoietic cells are closely related to the induction of oxidative stress in irradiated HSCs. The proton exposure-induced acute and long-term hematopoietic damage might be ameliorated through using antioxidants, which should be investigated in the future.

4. Oxygen radiation and hematopoietic stem cells

As we discussed above, GCR contains various HZE particles including ^{56}Fe, ^{28}Si, ^{16}O, and ^{12}C, which have more detrimental effects on normal tissues than do photon and proton radiations during spaceflight. Oxygen (^{16}O) radiation has relatively high-charge and high-linear energy transfer (LET), leading to a high relative biological effectiveness. In this section, we will mainly discuss the biological negative effects of ^{16}O on hematopoietic stem cells in long-duration space missions.

Hematopoietic cells in the body are the most radiosensitive cells to radiation [73, 74]. Exposure to γ-irradiation causes both acute and long-term damage in hematopoietic stem and progenitor cells (HSPCs), which is due primarily to radiation-induced cellular apoptosis and senescence in HSPCs [37–39, 65]. Using porcine and mice model, it has documented that proton radiation induced both acute and long-term hematopoietic damage. We have described the acute and residual effects of proton radiation on hematopoietic stem cells showing that numbers and function of bone marrow HSCs in mice were detrimentally affected. The negative effects of proton radiation mainly contribute to increasing the production of oxidative stress and DNA damage in irradiated HSCs [53]. ^{56}Fe radiation causes significant alterations in the expression of repetitive elements and DNA methylation, and 0.1-0.4 Gy of ^{56}Fe radiation resulted in significant epigenetic changes in hematopoietic stem and progenitor cells in a mouse model [75]. Using cultured human hematopoietic stem and progenitor cells, it was found that ^{12}C radiation induced chromosome aberrations and cellular apoptosis [76]. 0.3–0.9 Gy of ^{28}Si radiation triggers a significant increase of cellular apoptosis in irradiated mice HSCs at 4 weeks after the exposure, which results in the deficiency of numbers and clonogenic function of irradiated HSCs [77]. These findings indicate that GCR, including different forms of ionizing radiation, induces acute and residual injury in hematopoietic stem cells. However, it remains elusive whether ^{16}O radiation induces acute and long-term hematopoietic effects and what main factors are involved in the negative effects on HSCs under ^{16}O exposure.

In one of our experiments, C57BL/6 J mice were exposed to 0.1, 0.25, and 1.0 Gy ^{16}O (600 MeV/n) total-body irradiation (TBI) and analyzed the effects of ^{16}O radiation on peripheral blood and BM 2 weeks after the exposure [51] (**Table 3**). Since hematopoietic cells are known to be radiosensitive, it is not surprising that a significant decrease was observed in peripheral WBC and platelet counts in mice exposed to 1.0 Gy of ^{16}O. In comparison to ^{16}O radiation, peripheral blood cell counts, including numbers of WBCs and platelets, were almost recovered to normal levels at 2 weeks after 1.0 Gy of γ-ray radiation in BALB/c mice [78]. This might due to (1) different animal species used and (2) different biological effectiveness of ^{16}O and γ-ray radiation along with high LET properties of ^{16}O. ^{16}O TBI causes cellular apoptosis in hematopoietic progenitors but not hematopoietic stem cells at 2 weeks postexposure. To monitor how fast HPCs recover from ^{16}O TBI, apoptotic assay was performed at 3 months after 0.1, 0.25, and 1.0 Gy of ^{16}O TBI, showing that the apoptotic levels in HPCs and HSCs after the exposure are similar to those in nonirradiated mice. These data suggest that HPCs have a slower recovery than HSCs after ^{16}O TBI.

	Number of cells		Function of HSCs		
	HPC	LSK cells	BFU-E	CFU-GM	CFU-GEMM
2 weeks	↓	↓	↓	↓	↓
3 months	↓	↓	↓	↓	↓

Note: C57BL/6 mice were exposed to 1.0 Gy of oxygen total body irradiation. The indicated parameters were measured at 2 weeks and 3 months after the exposure. ↓ Indicate that the parameter is decreased when compared to non-irradiated mice.

Table 3.
Acute and long-term effects of oxygen irradiation on hematopoietic cells.

Previous studies have demonstrated that the functional defect of mouse HSCs was observed under either radiation or chemotherapeutic drug treatment. For example, the reconstitution capacity of mouse HSCs has detrimental effects under 1.0 Gy of total-body γ-ray radiation along with apparent myeloid bias differentiation [78, 79]. One dose of 5-fluorouracil treatment significantly decreased the numbers and engraftment ability of mouse HSCs at day 10 after the exposure [80]. To investigate the effects of low doses of ^{16}O TBI on HSC function, in vitro colony-forming assays using bone marrow cells were performed at 2 weeks after 0.1–1.0 Gy doses of ^{16}O exposure. It shows that low doses of ^{16}O TBI not only decreases numbers of HSCs but also abates the in vitro colony-forming abilities, such as decreased numbers of BFU-E, CFU-GM, and CFU-GEMM from irradiated bone marrow cells. These data suggest that the function of HSC after ^{16}O TBI was negative affected despite low doses of ^{16}O used. Using in vitro cell culture model, previous studies have shown that ^{16}O radiation has more dramatic effects on chromosomal aberrations, micronuclei formation, cell survival, and apoptosis than photon radiation [81, 82]. It has been documented that 6.5 Gy total-body γ-irradiation decreased the numbers of bone marrow HSCs up to 50% at 2 weeks postexposure, which was observed at 2 weeks after 1.0 Gy of total-body ^{16}O radiation [38]. These results indicate that ^{16}O radiation has a higher RBE than photon radiation.

To explore long-term effect of ^{16}O TBI on hematopoietic cells, 0.05, 0.1, 0.25, and 1.0 Gy ^{16}O (600 MeV/n) radiations were used to irradiate C57BL/6 J mice. Irradiated mice were analyzed for the long-term effects of ^{16}O radiation on peripheral blood cells and bone marrow cells 3 months postexposure [83]. Although there are the same numbers of peripheral blood cells at 3 months after 0.05 to 1.0 Gy of ^{16}O TBI as nonirradiated controls, numbers of HPC and HSCs from irradiated mice were significantly lower than those from nonirradiated controls. The changes of peripheral blood cell counts after oxygen radiation are similar to the effects of other types of ionizing radiation, such as 0.5 and 1.0 Gy of γ-TBI [84, 85]. Peripheral blood cell counts were back to normal levels 2 months after sublethal doses of γ-ray exposure [86]. Recovery of peripheral blood cell counts may neglect the effects of irradiation on bone marrow HSCs [84], which will result in overlooking the long-term bone marrow suppression after radiation.

We have demonstrated that 0.1 to 1.0 Gy of ^{16}O TBI, but not 0.05 Gy dose, resulted in a dramatic impairment in both numbers and function of bone marrow HSCs in mice at 3 months after exposure. Comparing to ^{16}O TBI, exposure of mice to 0.5 and 1.0 Gy of γ-TBI did not negatively affect the numbers and function of HSC in mice at 3 months after exposure. The phenotype from ^{16}O- and γ-TBI might be related to their RBE along with higher RBE levels of ^{16}O TBI than γ-TBI [81]. The long-term detrimental effects of ^{16}O TBI on bone marrow hematopoietic stem cells have also been seen in 6.5 Gy sublethal doses of γ-rays and 1.0 Gy low dose of proton radiation, showing a reduction in HSC reserves and a defect in HSC function [48, 85, 87, 88]. When comparing the effects of different radiation sources on HPCs, we have shown that acute exposure to low doses of ^{16}O TBI triggered a significant reduction in numbers of HPCs at 2 weeks after exposure. However, numbers of HPCs in irradiated mice recovered back to normal levels 2 weeks after either γ-ray or proton exposure [51]. These data indicate that ^{16}O-irradiated HPCs have a slower recovery than proton- and photon-irradiated HPCs.

We have previously demonstrated that exposure of mice to 1.0 Gy of ^{16}O TBI leads to an increased rate of apoptosis at 2 weeks postexposure in irradiated HPCs but not HSCs. This is consistent with HPC colony-forming ability assay, showing lower numbers of BFU-E, CFU-GM, and CFU-GEMM when compared to those in nonirradiated controls [51]. When we further examined HPC colony-forming abilities at 3 months after same dose of ^{16}O TBI, it showed that numbers of various colonies were still much lower than those in nonirradiated controls [83]. Notably, the decreased colony-forming abilities after ^{16}O TBI were in a dose-independent manner, which suggests "hit and damage." These data suggest that oxygen irradiation has features with a high-linear energy transfer and strong relative biological effectiveness.

A cobblestone area-forming cell (CAFC) assay is a surrogate in vitro hematopoietic stem cell functional assay. We measured HSC in vitro CAFC-forming ability at 3 months after 0.1, 0.25, and 1.0 Gy of ^{16}O radiation, showing that irradiated mice HSCs had a significant reduction of CAFC numbers independent of radiation doses (**Table 4**). These unusual dose-response curves have also been seen in the studies of ^{28}Si radiation, such as effects of ^{28}Si radiation on synaptic plasticity and contextual fear memory [89, 90]. However, when mice were exposed to 0.05 Gy of ^{16}O TBI, numbers of CAFC were comparable to nonirradiated controls, indicating that HSC function was not affected after exposure to 0.05 Gy ^{16}O TBI [83]. These long-term negative effects of ^{16}O TBI on HSCs are also observed in other different types of ionizing radiation [53, 63, 91]. Mice were exposed to 1.0 Gy of proton total-body irradiation, leading to a significant decrease in the CAFC-forming ability in HSCs at 22 weeks postexposure [53]. 6.5 Gy of γ-ray total-body irradiation caused a reduction in HSC colony-forming ability 2 months postexposure [38]. As we discussed previously, peripheral blood cell counts were back to normal levels at 3 months after low-dose ^{16}O TBI, while the numbers and function of bone marrow HSCs were significantly decreased after exposure. Proliferation of progenitors (such as myeloid, lymphoid, and erythroid progenitors) might contribute to the recovery of peripheral blood cells after radiation. These results suggest that low doses of ionizing radiation can induce long-term HSC suppression, while ^{16}O TBI has stronger abilities to induce the long-term HSC suppression than other types of ionizing radiation.

Cellular apoptosis and ROS production are crucial mediators in irradiation-induced cell damage. We exposed mice to 1.0 Gy of ^{16}O, and it triggered an aberrant increase in ROS production in HPCs and HSCs 2 weeks after exposure [51]. Meanwhile, increasing levels of apoptosis were significant in irradiated HPCs, but not LSK cells and HSCs, when compared to nonirradiated controls. Whether the different acute responses of HPCs and HSCs to ^{16}O TBI are related to ROS production has yet to be determined. Induction of ROS production persisted up to 3 months after ^{16}O TBI [83], which is congruent with the decreased expression of the antioxidant genes GPX2 and SOD3 in 1.0 Gy ^{16}O-irradiated HSCs when compared to nonirradiated HSCs. Previous studies utilized proton and γ-ray radiation to prove that HSC functional impairment might be attributable to the accumulation of residual ROS [53, 63, 92, 93]. This is evidenced by a decrease in in vivo

	HPC			LSK cells		
	Apoptosis	ROS	DNA damage	Apoptosis	ROS	DNA damage
2 weeks	↑	↑	no change	no change	↑	↑
3 months	no change	no change	no change	no change	↑	no change

Note: C57BL/6 mice were exposed to 1.0 Gy of oxygen total body irradiation. The indicated parameters were measured at 2 weeks and 3 months after the exposure.
↑ Indicate that the parameter is increased when compared to non-irradiated mice. No change means that the parameter is similar to that from non-irradiated mice.

Table 4.
Induction of cellular apoptosis, ROS, and DNA damage after oxygen irradiation.

engraftment capacity and in vitro colony-forming ability using bone marrow cells after radiation. The importance of ROS overproduction on functional HSCs was not only supported under radiation stress condition but also supported by other genetic animal models. Deletion of Foxo3, ATM, TSC1, and Bmi-1 in mice leads to the impairment of numbers and function of HSCs along with increasing ROS production. Application of antioxidants, such as N-acetyl cysteine (NAC), on these mutant mice significantly ameliorated the HSC functional deficiency [71, 94–96]. We have provided data showing ^{16}O-irradiated HSCs had higher levels of ROS production than nonirradiated animals in both acute and long-term studies. It is well accepted that mitochondrial oxidative phosphorylation and NADPH oxidases are two main sources to produce ROS in mammalian cells. Because HSCs reside in hypoxic environmental niche in the bone marrow and have higher expression of HIF1α in response to hypoxia, HSCs produce ROS mainly through glycolysis and NOX enzyme. We have previously shown that proton and γ-ray radiation induced significantly upregulation of NOX4 in irradiated HSCs [53, 66]. The NOX4 inhibitor diphenyliodonium can partially protect functional HSCs from γ-irradiation-induced long-term damage. Therefore, antioxidants, such as NOX4 inhibitors and NAC, should be further tested whether inhibiting ROS production can decrease ^{16}O TBI-induced ROS production to accelerate the functional recovery of HPCs and HSCs after ^{16}O irradiation exposure.

Under radiotherapy and chemotherapy stress conditions, dominant HSCs might be activated from quiescent status to provide the need for stressed hematopoietic system. However, frequent HSC activation might cause its loss of self-renewal ability, differentiation, and death with bone marrow failure syndrome [69, 97]. We have shown that proton and γ-irradiation can efficiently activate quiescent HSCs [53, 65], leading to the redistribution of different cell cycle phases and stem cell functional defects. Data from genetic mice models, such as depletion of FOXO3a and Lkb1, showed that HSCs had fast cycling with loss of HSC self-renewal ability and HSC exhaustion [94, 98–100]. There are fewer numbers of HSCs in G_0 and higher numbers in G_1/G_2SM at 2 weeks after ^{16}O TBI [51]. Additionally, we observed that around 15% of irradiated HSCs had more than two γH2AX foci per cell 2 weeks after ^{16}O exposure [51], which is positively correlated with the increased ROS production in ^{16}O TBI HSCs. Taken together, all of ROS production, DNA damage, and HSC cycling after ^{16}O TBI might contribute to HSC defect induced by oxygen radiation, which will be tested in our future studies.

Note:

1. Due to the large difference in size between mice and human subjects, the dose-depth distribution in mice exposed to protons is different from that in humans [1]. The dose to the blood-forming tissues in humans will be lower, whereas the dose to the blood-forming tissues in mice is likely very similar to the dose at the surface. This difference in dose-depth profile should be taken into considerations when translating risk of hematopoietic damage in response to space radiation in a mouse model to the human situation.

2. During long-term space flight (such as to Mars), astronauts will encounter either a chronic, fractionated dose of space radiation or they will accumulate 1–3 Gy of radiation exposure. Presently, it is impossible to achieve chronic or fractionated doses of space radiation due to the limited access to the facilities capable of producing synthetic space radiation. In most research to date, investigators have used either acute radiation exposure or higher doses of space radiation in animal models. Cautions should therefore be taken when applying those findings from animal studies to human.

3. The protons and oxygen nuclei in the studies described here were all delivered within a few minutes, and most charged particle exposures during space flight occur at a very low-dose rate and/or are fractionated. Though we acknowledge that the high-dose rates we used are a limitation of our studies, low-dose rate exposures were not possible because of practical constraints. Future studies with low-dose rate or fractionated exposures should provide further insight into dose-rate dependence of hematopoietic stem/progenitor cell response to high-energy charged particle radiation.

5. Conclusion

In summary, proton and oxygen space radiation have detrimental effects on the hematopoietic system even with at low doses, which will have potential implications for health outcomes during long-duration space missions. Increasing ROS production might be a major mediator on space radiation-induced HSC damage. Knowledge gained from this chapter could aid in planning countermeasure strategies to protect against hematopoietic effects of radiation exposure during space travel. To minimize the health negative effects of deep-space travel, decreasing oxidative stress might be a good approach to mitigate the adverse effects of proton and HZE particle exposure on the hematopoietic system.

Acknowledgements

The study was in part supported by the National Natural Science Foundation of China (Grant No. 81460110, 81660123, and 81860026),

Conflict of interest

The authors declare no conflicts of interest.

Author details

Huihong Zeng[1], Mengzhen Yue[1] and Lijian Shao[1,2]*

1 Medical College of Nanchang University, Nanchang, China

2 Jiangxi Provincial Key Laboratory of Preventive Medicine, Nanchang University, Nanchang, China

*Address all correspondence to: lshao@ncu.edu.cn

References

[1] Cengel KA, Diffenderfer ES, Avery S, Kennedy AR, McDonough J. Using electron beam radiation to simulate the dose distribution for whole body solar particle event proton exposure. Radiation and Environmental Biophysics. 2010;**49**(4):715-721

[2] Townsend LW. Implications of the space radiation environment for human exploration in deep space. Radiation Protection Dosimetry. 2005;**115**(1-4):44-50

[3] Simonsen LC, Cucinotta FA, Atwell W, Nealy JE. Temporal analysis of the October 1989 proton flare using computerized anatomical models. Radiation Research. 1993;**133**(1):1-11

[4] Parsons JL, Townsend LW. Interplanetary crew dose rates for the August 1972 solar particle event. Radiation Research. 2000;**153**(6):729-733

[5] Moore FD. Radiation burdens for humans on prolonged exomagnetospheric voyages. FASEB Journal: Official Publication of the Federation of American Societies for Experimental Biology. 1992;**6**(6):2338-2343

[6] Datta K, Suman S, Trani D, Doiron K, Rotolo JA, Kallakury BV, et al. Accelerated hematopoietic toxicity by high energy (56)Fe radiation. International Journal of Radiation Biology. 2012;**88**(3):213-222

[7] Boerma M, Sridharan V, Mao XW, Nelson GA, Cheema AK, Koturbash I, et al. Effects of ionizing radiation on the heart. Mutation Research. 2016;**770**(Pt B):319-327

[8] Suman S, Datta K, Trani D, Laiakis EC, Strawn SJ, Fornace AJ Jr. Relative biological effectiveness of 12C and 28Si radiation in C57BL/6J mice. Radiation and Environmental Biophysics. 2012;**51**(3):303-309

[9] Cucinotta FA, Wu H, Shavers MR, George K. Radiation dosimetry and biophysical models of space radiation effects. Gravitational and Space Biology Bulletin. 2003;**16**(2):11-18

[10] Walker SA, Townsend LW, Norbury JW. Heavy ion contributions to organ dose equivalent for the 1977 galactic cosmic ray spectrum. Advances in Space Research. 2013;**51**(9):1792-1799

[11] Zeitlin C, Hassler DM, Cucinotta FA, Ehresmann B, Wimmer-Schweingruber RF, Brinza DE, et al. Measurements of energetic particle radiation in transit to Mars on the Mars Science Laboratory. Science. 2013;**340**(6136):1080-1084

[12] Hamilton SA, Pecaut MJ, Gridley DS, Travis ND, Bandstra ER, Willey JS, et al. A murine model for bone loss from therapeutic and space-relevant sources of radiation. Journal of Applied Physiology. 2006;**101**(3):789-793

[13] Shouse SS, Warren SL, Whipple GH. II. Aplasia of marrow and fatal intoxication in dogs produced by roentgen radiation of all bones. The Journal of Experimental Medicine. 1931;**53**(3):421-435

[14] Raber J, Yamazaki J, Torres ERS, Kirchoff N, Stagaman K, Sharpton T, et al. Combined effects of three high-energy charged particle beams important for space flight on brain, behavioral and cognitive endpoints in B6D2F1 female and male mice. Frontiers in Physiology. 2019;**10**:179

[15] Iancu OD, Boutros SW, Olsen RHJ, Davis MJ, Stewart B, Eiwaz M, et al. Space radiation alters

genotype-phenotype correlations in fear learning and memory tests. Frontiers in Genetics. 2018;**9**:404

[16] Warren SL, Whipple GH. Roentgen ray intoxication: I. Unit dose over thorax negative-over abdomen lethal. Epithelium of small intestine sensitive to X-rays. The Journal of Experimental Medicine. 1922;**35**(2):187-202

[17] Jacobson LO. Evidence for a humoral factor (or factors) concerned in recovery from radiation injury: A review. Cancer Research. 1952;**12**(5):315-325

[18] Jacobson LO. Hematopoietic responses to radiation injury. Annual Review of Medicine. 1956;**7**:345-352

[19] Ford CE, Hamerton JL, Barnes DW, Loutit JF. Cytological identification of radiation-chimaeras. Nature. 1956;**177**(4506):452-454

[20] Nowell PC, Cole LJ, Habermeyer JG, Roan PL. Growth and continued function of rat marrow cells in x-radiated mice. Cancer Research. 1956;**16**(3):258-261

[21] Trentin JJ. Mortality and skin transplantability in x-irradiated mice receiving isologous, homologous or heterologous bone marrow. Proceedings of the Society for Experimental Biology and Medicine. 1956;**92**(4):688-693

[22] Becker AJ, Mc CE, Till JE. Cytological demonstration of the clonal nature of spleen colonies derived from transplanted mouse marrow cells. Nature. 1963;**197**:452-454

[23] Till JE, Mc CE. A direct measurement of the radiation sensitivity of normal mouse bone marrow cells. Radiation Research. 1961;**14**:213-222

[24] Lorimore SA, Coates PJ, Wright EG. Radiation-induced genomic instability and bystander effects: Inter-related nontargeted effects of exposure to ionizing radiation. Oncogene. 2003;**22**(45):7058-7069

[25] Morgan WF. Non-targeted and delayed effects of exposure to ionizing radiation: II. Radiation-induced genomic instability and bystander effects in vivo, clastogenic factors and transgenerational effects. Radiation Research. 2003;**159**(5):581-596

[26] McCulloch EA, Till JE. The radiation sensitivity of normal mouse bone marrow cells, determined by quantitative marrow transplantation into irradiated mice. Radiation Research. 1960;**13**:115-125

[27] Kiel MJ, Yilmaz OH, Iwashita T, Yilmaz OH, Terhorst C, Morrison SJ. SLAM family receptors distinguish hematopoietic stem and progenitor cells and reveal endothelial niches for stem cells. Cell. 2005;**121**(7):1109-1121

[28] Goodell MA, Brose K, Paradis G, Conner AS, Mulligan RC. Isolation and functional properties of murine hematopoietic stem cells that are replicating in vivo. The Journal of Experimental Medicine. 1996;**183**(4):1797-1806

[29] Weissman IL, Anderson DJ, Gage F. Stem and progenitor cells: Origins, phenotypes, lineage commitments, and transdifferentiations. Annual Review of Cell and Developmental Biology. 2001;**17**:387-403

[30] Wilson A, Laurenti E, Trumpp A. Balancing dormant and self-renewing hematopoietic stem cells. Current Opinion in Genetics & Development. 2009;**19**(5):461-468

[31] Wang Y, Schulte BA, Zhou D. Hematopoietic stem cell senescence and long-term bone marrow injury. Cell Cycle. 2006;**5**(1):35-38

[32] Nelson GA. Space radiation and human exposures, a primer. Radiation Research. 2016;**185**(4):349-358. Epub 2016/03/29

[33] Raju MR. Proton radiobiology, radiosurgery and radiotherapy. International Journal of Radiation Biology. 1995;**67**(3):237-259

[34] Loeffler JS, Smith AR, Suit HD. The potential role of proton beams in radiation oncology. Seminars in Oncology. 1997;**24**(6):686-695

[35] Chambers KA, Harrington NP, Ross WM, Filion LG. Relative alterations in blood mononuclear cell populations reflect radiation injury in mice. Cytometry. 1998;**31**(1):45-52

[36] Harrington NP, Chambers KA, Ross WM, Filion LG. Radiation damage and immune suppression in splenic mononuclear cell populations. Clinical and Experimental Immunology. 1997;**107**(2):417-424

[37] Shao L, Sun Y, Zhang Z, Feng W, Gao Y, Cai Z, et al. Deletion of proapoptotic Puma selectively protects hematopoietic stem and progenitor cells against high-dose radiation. Blood. 2010;**115**(23):4707-4714

[38] Wang Y, Schulte BA, LaRue AC, Ogawa M, Zhou D. Total body irradiation selectively induces murine hematopoietic stem cell senescence. Blood. 2006;**107**(1):358-366

[39] Yu H, Shen H, Yuan Y, XuFeng R, Hu X, Garrison SP, et al. Deletion of Puma protects hematopoietic stem cells and confers long-term survival in response to high-dose gamma-irradiation. Blood. 2010;**115**(17):3472-3480

[40] Kajioka EH, Andres ML, Li J, Mao XW, Moyers MF, Nelson GA, et al. Acute effects of whole-body proton irradiation on the immune system of the mouse. Radiation Research. 2000;**153**(5 Pt 1):587-594

[41] Gridley DS, Rizvi A, Luo-Owen X, Makinde AY, Coutrakon GB, Koss P, et al. Variable hematopoietic responses to acute photons, protons and simulated solar particle event protons. In Vivo. 2008;**22**(2):159-169

[42] Ware JH, Sanzari J, Avery S, Sayers C, Krigsfeld G, Nuth M, et al. Effects of proton radiation dose, dose rate and dose fractionation on hematopoietic cells in mice. Radiation Research. 2010;**174**(3):325-330

[43] Maks CJ, Wan XS, Ware JH, Romero-Weaver AL, Sanzari JK, Wilson JM, et al. Analysis of white blood cell counts in mice after gamma- or proton-radiation exposure. Radiation Research. 2011;**176**(2):170-176

[44] Luo-Owen X, Pecaut MJ, Rizvi A, Gridley DS. Low-dose total-body gamma irradiation modulates immune response to acute proton radiation. Radiation Research. 2012;**177**(3):251-264

[45] Sanzari JK, Cengel KA, Steven Wan X, Rusek A, Kennedy AR. Acute hematological effects in mice exposed to the expected doses, dose-rates, and energies of solar particle event-like proton radiation. Life Sciences and Space Research. 2014;**2**:86-91

[46] Wagemaker G. Heterogeneity of radiation sensitivity of hemopoietic stem cell subsets. Stem Cells. 1995;**13**(Suppl 1):257-260

[47] Gridley DS, Pecaut MJ. Whole-body irradiation and long-term modification of bone marrow-derived cell populations by low- and high-LET radiation. In Vivo. 2006;**20**(6B):781-789

[48] Shao L, Luo Y, Zhou D. Hematopoietic stem cell injury induced by ionizing radiation. Antioxidants & Redox Signaling. 2014;**20**(9):1447-1462

[49] Shao L, Wang Y, Chang J, Luo Y, Meng A, Zhou D. Hematopoietic stem cell senescence and cancer therapy-induced long-term bone marrow injury. Translational Cancer Research. 2013;**2**(5):397-411

[50] Chang J, Wang Y, Pathak R, Sridharan V, Jones T, Mao XW, et al. Whole body proton irradiation causes acute damage to bone marrow hematopoietic progenitor and stem cells in mice. International Journal of Radiation Biology. 2017;**93**(12):1312-1320

[51] Chang J, Luo Y, Wang Y, Pathak R, Sridharan V, Jones T, et al. Low doses of oxygen ion irradiation cause acute damage to hematopoietic cells in mice. PLoS One. 2016;**11**(7):e0158097

[52] Cucinotta FA, Plante I, Ponomarev AL, Kim MH. Nuclear interactions in heavy ion transport and event-based risk models. Radiation Protection Dosimetry. 2011;**143**(2-4):384-390

[53] Chang J, Feng W, Wang Y, Luo Y, Allen AR, Koturbash I, et al. Whole-body proton irradiation causes long term damage to hematopoietic stem cells in mice. Radiation Research. 2015;**183**(2):240-248

[54] Major IR, Mole RH. Myeloid leukaemia in x-ray irradiated CBA mice. Nature. 1978;**272**(5652):455-456

[55] Shimizu Y, Kato H, Schull WJ. Risk of cancer among atomic bomb survivors. Journal of Radiation Research. 1991;**32**(Suppl 2):54-63

[56] Kennedy AR, Davis JG, Carlton W, Ware JH. Effects of dietary antioxidant supplementation on the development of malignant lymphoma and other neoplastic lesions in mice exposed to proton or iron-ion radiation. Radiation Research. 2008;**169**(6):615-625

[57] Sanzari JK, Wan SX, Diffenderfer ES, Cengel KA, Kennedy AR. Relative biological effectiveness of simulated solar particle event proton radiation to induce acute hematological change in the porcine model. Journal of Radiation Research. 2014;**55**(2):228-244

[58] Desai N, Davis E, O'Neill P, Durante M, Cucinotta FA, Wu H. Immunofluorescence detection of clustered gamma-H2AX foci induced by HZE-particle radiation. Radiation Research. 2005;**164**(4 Pt 2):518-522

[59] Goetz W, Morgan MN, Baulch JE. The effect of radiation quality on genomic DNA methylation profiles in irradiated human cell lines. Radiation Research. 2011;**175**(5):575-587

[60] Tian J, Zhao W, Tian S, Slater JM, Deng Z, Gridley DS. Expression of genes involved in mouse lung cell differentiation/regulation after acute exposure to photons and protons with or without low-dose preirradiation. Radiation Research. 2011;**176**(5):553-564

[61] Di Pietro C, Piro S, Tabbi G, Ragusa M, Di Pietro V, Zimmitti V, et al. Cellular and molecular effects of protons: Apoptosis induction and potential implications for cancer therapy. Apoptosis. 2006;**11**(1):57-66

[62] Dorr H, Meineke V. Acute radiation syndrome caused by accidental radiation exposure - therapeutic principles. BMC Medicine. 2011;**9**:126

[63] Li H, Wang Y, Pazhanisamy SK, Shao L, Batinic-Haberle I, Meng A, et al. Mn(III) meso-tetrakis-(N-ethylpyridinium-2-yl) porphyrin mitigates total body irradiation-induced long-term bone marrow suppression. Free Radical Biology & Medicine. 2011;**51**(1):30-37

[64] Zhang H, Zhai Z, Wang Y, Zhang J, Wu H, Li C, et al. Resveratrol

ameliorates ionizing irradiation-induced long-term hematopoietic stem cell injury in mice. Free Radical Biology & Medicine. 2013;**54**:40-50

[65] Shao L, Feng W, Li H, Gardner D, Luo Y, Wang Y, et al. Total body irradiation causes long-term mouse BM injury via induction of HSC premature senescence in an Ink4a- and Arf-independent manner. Blood. 2014;**123**(20):3105-3115

[66] Wang Y, Liu L, Pazhanisamy SK, Li H, Meng A, Zhou D. Total body irradiation causes residual bone marrow injury by induction of persistent oxidative stress in murine hematopoietic stem cells. Free Radical Biology & Medicine. 2010;**48**(2):348-356

[67] Takubo K, Goda N, Yamada W, Iriuchishima H, Ikeda E, Kubota Y, et al. Regulation of the HIF-1alpha level is essential for hematopoietic stem cells. Cell Stem Cell. 2010;**7**(3):391-402

[68] Pazhanisamy SK, Li H, Wang Y, Batinic-Haberle I, Zhou D. NADPH oxidase inhibition attenuates total body irradiation-induced haematopoietic genomic instability. Mutagenesis. 2011;**26**(3):431-435

[69] Wilson A, Laurenti E, Oser G, van der Wath RC, Blanco-Bose W, Jaworski M, et al. Hematopoietic stem cells reversibly switch from dormancy to self-renewal during homeostasis and repair. Cell. 2008;**135**(6):1118-1129

[70] Tothova Z, Kollipara R, Huntly BJ, Lee BH, Castrillon DH, Cullen DE, et al. FoxOs are critical mediators of hematopoietic stem cell resistance to physiologic oxidative stress. Cell. 2007;**128**(2):325-339

[71] Chen C, Liu Y, Liu R, Ikenoue T, Guan KL, Liu Y, et al. TSC-mTOR maintains quiescence and function of hematopoietic stem cells by repressing mitochondrial biogenesis and reactive oxygen species. The Journal of Experimental Medicine. 2008;**205**(10):2397-2408

[72] Schuringa JJ, Vellenga E. Role of the polycomb group gene BMI1 in normal and leukemic hematopoietic stem and progenitor cells. Current Opinion in Hematology. 2010;**17**(4):294-299

[73] Waselenko JK, MacVittie TJ, Blakely WF, Pesik N, Wiley AL, Dickerson WE, et al. Medical management of the acute radiation syndrome: Recommendations of the Strategic National Stockpile Radiation Working Group. Annals of Internal Medicine. 2004;**140**(12):1037-1051

[74] Moroni M, Coolbaugh TV, Lombardini E, Mitchell JM, Moccia KD, Shelton LJ, et al. Hematopoietic radiation syndrome in the Gottingen minipig. Radiation Research. 2011;**176**(1):89-101

[75] Miousse IR, Shao L, Chang J, Feng W, Wang Y, Allen AR, et al. Exposure to low-dose (56)Fe-ion radiation induces long-term epigenetic alterations in mouse bone marrow hematopoietic progenitor and stem cells. Radiation Research. 2014;**182**(1):92-101

[76] Becker D, Elsasser T, Tonn T, Seifried E, Durante M, Ritter S, et al. Response of human hematopoietic stem and progenitor cells to energetic carbon ions. International Journal of Radiation Biology. 2009;**85**(11):1051-1059

[77] Chang J, Feng W, Wang Y, Allen AR, Turner J, Stewart B, et al. (28)Si total body irradiation injures bone marrow hematopoietic stem cells via induction of cellular apoptosis. Life Sciences and Space Research. 2017;**13**:39-44

[78] Stewart FM, Zhong S, Wuu J, Hsieh C, Nilsson SK, Quesenberry PJ. Lymphohematopoietic engraftment in minimally myeloablated hosts. Blood. 1998;**91**(10):3681-3687

[79] Stewart FM, Zhong S, Lambert JF, Colvin GA, Abedi M, Dooner MS, et al. Host marrow stem cell potential and engraftability at varying times after low-dose whole-body irradiation. Blood. 2001;**98**(4):1246-1251

[80] Haug JS, He XC, Grindley JC, Wunderlich JP, Gaudenz K, Ross JT, et al. N-cadherin expression level distinguishes reserved versus primed states of hematopoietic stem cells. Cell Stem Cell. 2008;**2**(4):367-379

[81] Habermehl D, Ilicic K, Dehne S, Rieken S, Orschiedt L, Brons S, et al. The relative biological effectiveness for carbon and oxygen ion beams using the raster-scanning technique in hepatocellular carcinoma cell lines. PLoS One. 2014;**9**(12):e113591

[82] Pathak R, Dey SK, Sarma A, Khuda-Bukhsh AR. Cell killing, nuclear damage and apoptosis in Chinese hamster V79 cells after irradiation with heavy-ion beams of (16)O, (12) C and (7)Li. Mutation Research. 2007;**632**(1-2):58-68

[83] Wang Y, Chang J, Li X, Pathak R, Sridharan V, Jones T, et al. Low doses of oxygen ion irradiation cause long-term damage to bone marrow hematopoietic progenitor and stem cells in mice. PLoS One. 2017;**12**(12):e0189466

[84] Chang J, Wang Y, Shao L, Laberge RM, Demaria M, Campisi J, et al. Clearance of senescent cells by ABT263 rejuvenates aged hematopoietic stem cells in mice. Nature Medicine. 2016;**22**(1):78-83

[85] Xu G, Wu H, Zhang J, Li D, Wang Y, Zhang H, et al. Metformin ameliorates ionizing irradiation-induced long-term hematopoietic stem cell injury in mice. Free Radical Biology & Medicine. 2015;**87**:15-25

[86] Testa NG, Hendry JH, Molineux G. Long-term bone marrow damage in experimental systems and in patients after radiation or chemotherapy. Anticancer Research. 1985;**5**(1):101-110

[87] Farres J, Martin-Caballero J, Martinez C, Lozano JJ, Llacuna L, Ampurdanes C, et al. Parp-2 is required to maintain hematopoiesis following sublethal gamma-irradiation in mice. Blood. 2013;**122**(1):44-54

[88] Chua HL, Plett PA, Sampson CH, Joshi M, Tabbey R, Katz BP, et al. Long-term hematopoietic stem cell damage in a murine model of the hematopoietic syndrome of the acute radiation syndrome. Health Physics. 2012;**103**(4):356-366

[89] Raber J, Marzulla T, Stewart B, Kronenberg A, Turker MS. 28Silicon irradiation impairs contextual fear memory in B6D2F1 mice. Radiation Research. 2015;**183**(6):708-712

[90] Raber J, Rudobeck E, Campbell-Beachler M, Allen AR, Allen B, Rosi S, et al. (28)Silicon radiation-induced enhancement of synaptic plasticity in the hippocampus of naive and cognitively tested mice. Radiation Research. 2014;**181**(4):362-368

[91] Epperly MW, Wang H, Jones JA, Dixon T, Montesinos CA, Greenberger JS. Antioxidant-chemoprevention diet ameliorates late effects of total-body irradiation and supplements radioprotection by MnSOD-plasmid liposome administration. Radiation Research. 2011;**175**(6):759-765

[92] Chai X, Li D, Cao X, Zhang Y, Mu J, Lu W, et al. ROS-mediated iron overload injures the hematopoiesis of bone marrow by damaging hematopoietic stem/progenitor cells in mice. Scientific Reports. 2015;**5**:10181

[93] Cho J, Yusuf R, Kook S, Attar E, Lee D, Park B, et al. Purinergic P2Y(1)(4) receptor modulates stress-induced

hematopoietic stem/progenitor cell senescence. The Journal of Clinical Investigation. 2014;**124**(7):3159-3171

[94] Miyamoto K, Araki KY, Naka K, Arai F, Takubo K, Yamazaki S, et al. Foxo3a is essential for maintenance of the hematopoietic stem cell pool. Cell Stem Cell. 2007;**1**(1):101-112

[95] Nitta E, Yamashita M, Hosokawa K, Xian M, Takubo K, Arai F, et al. Telomerase reverse transcriptase protects ATM-deficient hematopoietic stem cells from ROS-induced apoptosis through a telomere-independent mechanism. Blood. 2011;**117**(16):4169-4180

[96] Nakamura S, Oshima M, Yuan J, Saraya A, Miyagi S, Konuma T, et al. Bmi1 confers resistance to oxidative stress on hematopoietic stem cells. PLoS One. 2012;7(5):e36209

[97] Pietras EM, Warr MR, Passegue E. Cell cycle regulation in hematopoietic stem cells. The Journal of Cell Biology. 2011;**195**(5):709-720

[98] Nakada D, Saunders TL, Morrison SJ. Lkb1 regulates cell cycle and energy metabolism in haematopoietic stem cells. Nature. 2010;**468**(7324):653-658

[99] Lorenz E, Uphoff D, Reid TR, Shelton E. Modification of irradiation injury in mice and guinea pigs by bone marrow injections. Journal of the National Cancer Institute. 1951;**12**:197-201

[100] Main JM, Prehn RT. Successful skin homografts after the administration of high dosage X radiation and homologous bone marrow. Journal of the National Cancer Institute. 1955;**15**:1023-1029